MAN AND BEAST

MAN AND BEAST

by

Major C. W. HUME, OBE, MC, BSc.

Founder of UFAW (The Universities Federation for Animal Welfare); Citoyen d'Honneur de Meurchin (P. de C.); Membre d'Ounour dou Riban d'Arle (B. du R.); Albert Schweitzer Medallist 1956.

with drawings by
FOUGASSE

MEMORIAL EDITION

THE UNIVERSITIES FEDERATION FOR ANIMAL WELFARE
8 Hamilton Close, South Mimms, Potters Bar, Herts EN6 3QD
England

First published December 1962
Reprinted in paperback as the Memorial Edition 1982

The first edition was dedicated to
MARGARET PATTISON, M.A. (St. Andrews)
who, in 1966, became
MRS CHARLES HUME

ISBN 0 900767 28 6

Printed in Great Britain by
The Garden City Press Limited
Letchworth, Hertfordshire SG6 1JS

Charles Westley Hume, OBE MC BSc
1886-1981

This book, the second edition of *Man and Beast*, is dedicated to the memory of UFAW's Founder, Major Hume. The essays herein were written in the years 1943 to 1961; they are a record of his unique contribution to the cause of animal welfare and show how he aimed to promote humane behaviour towards all animals, including the unpopular ones.

Readers will note that in these days when Animals' Rights have become a rallying point for many animal lovers, the earliest essay, *Expanding Justice*, discussed the legal rights of animals, and the latest of them is entitled *What Rights have Animals?* His philosophy of animal welfare combined science with ethics and it is in the conviction that it is as relevant as ever today that we offer this book, unchanged in content, to a new generation. We have been encouraged by many of Major Hume's friends and admirers, in particular Mrs Christine Stevens of the Animal Welfare Institute, Washington DC, who in the early 1960s worked closely with him in promoting animal welfare legislation in the USA.

CONTENTS

INTRODUCTION

Those who work for UFAW in various capacities believe that a book that deals authoritatively with both the ethics of humanitarian endeavour and its practical expression is likely to be of considerable service to the layman as well as to the animal-welfare worker. They also believe that it is a great pity that many of the writings of Major Hume, UFAW's founder—especially those which deal with the ethical and philosophical aspects of our relations with animals, and those which describe the practical progress of animal-welfare campaigns from their small beginnings to their final successes—should virtually be lost in the back numbers of the periodicals, such as *Nature*, *The Lancet*, *The Hibbert Journal* and *The UFAW Courier*, in which they were originally published.

Thus there is a double reason for the publication of this book. Moreover, there is another incentive, for, if one reads between the lines as well as on them, one gets an encouraging picture of the change that has come about during the last thirty years or so in the general attitude towards animal welfare, and an equally encouraging picture of the way in which reforms which seem at first impossible to accomplish can be brought about with patience, tact, perseverance and the capacity to see and understand the point of view of other people—especially when these qualities are combined with knowledge, wisdom, faith and imperviousness to ridicule.

The book also shows, we think, how much of all these qualities of both heart and head the writer of the following articles possesses, and how well he combines ethical and philosophical reasoning with the practical application of scientific knowledge. It is, indeed, in the application of scientific knowledge that he has made his most distinctive contribution to animal welfare.

UFAW—or, to give it its full title, the Universities Federation for Animal Welfare—was founded in 1926 as the University of London Animal Welfare Society (ULAWS) as a result of a meeting at Birkbeck College, arranged by Major Hume, at which he was commissioned to form a university society. He had for a long time believed that animal problems must be tackled on a scientific basis, with a maximum of sympathy but a minimum of sentimentality, and, as a scientist and university graduate, he realized that the universities provided the best hunting-grounds for the right sort of personnel, both scientific and other, and also for the right sort of information to help with animal-welfare activities.

When he undertook the formation of ULAWS, he was already an Examiner in the Patent Office, Hon. Secretary of the British Science Guild (which stemmed from the British Association for the Advancement of Science and is now re-absorbed into it), Editor to the Physical Society, Brigade Signal Officer of the 142 Infantry Brigade (T.A.), and engaged in numerous other activities. Nevertheless, just as Satan finds employment for idle hands, so Heaven always seems to find still more employment for hands that are already more than fully occupied, and so Major Hume started ULAWS, with an initial membership of two.

It took seven years to bring the Society's annual income up to £100, but it then grew rapidly, and by 1938 it had spread to a number of other universities and its title had become UFAW. After the hazards of the Second World War (during which Major Hume returned to the Army) the new Federation forged ahead, and it is now held in great esteem and has considerable influence, especially in scientific circles, both in this country and internationally.

Throughout, Major Hume has been the leading spirit in the organization he founded. In addition to much other active work on its behalf, he has through the years produced many articles, pamphlets, leaflets and letters to the Press, and in 1956 he wrote *The Status of Animals in the Christian Religion*. In the same year he was awarded the Schweitzer Medal of the Animal Welfare Institute of America and, in 1962, the O.B.E., distinctions which gave all his colleagues a great deal of pleasure.

The following fifteen articles deal with the theory and practice of the chief campaigns with which he and UFAW have been concerned, with the exception of the first article, *The Principles of Animal Protection : The Philosophy of UFAW*, which was written specially for this volume : they are not in chronological order, but grouped as seemed most appropriate.

No. 3, *The Gin Trap : UFAW's Long Battle* (which might more justly be called "Major Hume's Long Battle"), describes one of his first campaigns, from its small start in 1928 up to its ultimate achievement in the banning of the Gin Trap in England and Wales as from August 1, 1958, and provides perhaps the best example of patience and perseverance and the refusal to be discouraged by almost insuperable difficulties. No. 7, *Electrocution : A Historical Retrospect*, describes another fight which took almost as long, and in which UFAW stood practically alone so far as England and Wales were concerned. No. 15, *Expanding Justice*, makes a fitting finale, summarizing the many problems which Major Hume has tackled and the great advance which UFAW (and also animal welfare in general, and the public attitude towards it) has made since it was founded as ULAWS in 1926.

1st September, 1962.

The Principles of Animal-Protection : The Philosophy of UFAW

CHARITY IS indivisible. The movement for protecting animals from cruelty in the first stage, and for positively promoting their welfare in the second stage, has been of a piece with the general movement for social reform which has characterized the past two centuries. Concern for the welfare of human beings and a similar concern for that of animals have been interlinked in history, and it is a fact of experience that those who care most sincerely for either of these good causes care also for the other.

All good causes have their fanatics, who may do some good but also inadvertently do harm to the enterprise that they espouse. The heart has its part to play in motivating social reform, but the head plays an equally important part in devising effective means for achieving that purpose, and UFAW came into existence in order to focus attention on the importance of authentic knowledge and accurate thought for effectively befriending animals. Throughout UFAW's history, realism, as opposed to sentimentality, has been the keynote of its policy.

The main requirement for realistic good will towards animals is a determination to improve their lot in actual fact. The enunciation of impracticable ideals may gratify the feelings of high-minded human beings but does not, in fact, ease the feelings of suffering animals. The distinction can best be explained by means of examples.

In the campaign against the iniquitous gin trap one of our most valuable allies was the Master of the Carmarthen Hounds. But although he was playing an indispensable part in the attack on the gin, and the final prohibition of that instrument of torture in 1958 could scarcely have been achieved without his help, UFAW was strongly criticized for associating with a Master of Hounds, and an attempt was made to deprive it of funds because it persisted in maintaining that association. At the time between thirty and forty million animals per annum were having their limbs crushed by the steel jaws of the trap and were being

left to suffer all night or longer, and our critics felt it better that this should continue rather than that we should soil our hands by associating with a hunting man. They would not, of course, have put it that way, but that is the practical effect which this abstract idealism would have had on the animals concerned.

Again, it often happens that the kindest thing to do to an animal is to kill it painlessly, but many people feel great distress if they have to do this, and they prefer to let it suffer. A good many continental animal-welfare societies collect stray cats and dogs off the streets so far as their funds permit, and keep them alive indefinitely at great expense, but will not kill those for which no homes can be found. In this way the number of animals they can befriend and the scope of the work they can do are unnecessarily restricted. In India many Hindus are unwilling to kill cows which have ceased to be useful, but turn them out to starve instead of putting them out of their misery. Many French residents leaving Algeria have not had the heart to kill their pets, but have chosen rather to leave them to die of neglect and starvation.

All these examples illustrate the unconscious admixture of selfishness with a love of animals which gives rise to unrealistic policies. Another outstanding instance is the tendency to care only about those species which one happens to know and understand best. Such preferences are accidental and have little to do with merit. The ancient Jews particularly loved sheep, the ancient Egyptians cats, the Hindus cows, and the Victorian English horses, which are protected above all other species by the Cruelty to Animals Act of 1876. The Home Office report on Cruelty to Wild Animals (Cmd. 8266, sec. 19, June, 1951) noted that :—

> "The sentimental concern about animals to which we have referred is directed mainly towards particular animals, such as foxes, deer and rabbits, which are, generally speaking, beautiful or attractive creatures. . . . Few people seem to be in the least concerned about what happens to rats. . . . Yet the rat is an intelligent and highly sensitive creature and probably suffers far more than some of the animals which attract a great deal of sentimental interest."

Only by clear thinking and objective assessment of the facts can all this selfishness be eliminated.

Fact-finding is an indispensable basis for a sound policy,

and here UFAW's close association with the scientific, veterinary, and medical worlds has been of value. Clearly this is true in respect of facts which depend on scientific research, as in the case of electrocution where UFAW was able to prevent an immense amount of suffering; and research has become an essential part of UFAW's practice. But our scientific association has, I think, served a wider purpose by safeguarding us against a trap into which policy-makers all too often fall, that of first choosing a policy and then making out a case for it, instead of first finding the facts and then basing a policy on them. For example, the horror felt over experiments on animals has led to the extreme policy of anti-vivisection, but a true picture of the relevant facts will be discriminatory, and in order to achieve discrimination a wide and accurate knowledge of very technical subject-matter is needed. Lacking this, too many controversialists have adopted an unrealistic policy and made out a case for it by means of sweeping generalizations which are untenable. Then again the conditions under which performing animals are trained and kept may be open to criticism, but propaganda on the subject has been so undiscriminating, and the true facts are so difficult to assess, that the issue has been deadlocked for some forty years.

We have also tried from the beginning to avoid another pit into which policy-makers can easily fall, and that is the shaping of policy with reference to its effect on popularity. All charitable societies depend on voluntary subscriptions and legacies, and popularity is a potent factor in obtaining these. On the other hand the welfare of animals depends on factors lying beyond the ken of many animal-lovers, who are more often sentimental than humane, so that a bid for popularity may run counter to that welfare. From the start we have tried to decide, in the light of our special knowledge, what policy will best help animals, and then to promote it without any regard to its popularity or unpopularity. As a result we have built up our support from among people who accept this way of thinking, and so we are not troubled by cranky pressure groups.

Here is a further consideration relating to policy-making. Human judgements fall into three classes—judgements of fact, judgements of moral principle, and judgements of aesthetic value. If I find that bullfighting entails the bruising and occasional evisceration of horses and the slow killing of bulls, that is a judgement of fact; if I find that it is painful but highly artistic, that is an aesthetic judgement; and if I find that it is evil, that is a moral judgement. Plato confused these three kinds of judgement by making them dependent on one another, and his example has too often been followed, but one of the greatest services rendered by physical science has been to show that sound conclusions cannot be reached unless judgements of each kind are made in complete independence of each of the other kinds. For instance, the question whether an allegation is true or false has to be isolated from the question whether it is congenial or distasteful, edifying or shocking, favourable or unfavourable to the policy that is being advocated.

A scale of priorities is necessary, and in assessing the relative importance of any practice which may cause suffering to animals UFAW has tried to be guided by the following considerations and no others:—

(1) the intensity of the suffering involved;
(2) its usual duration;
(3) the number of animals affected; and
(4) the feasibility of practical reform.

Thus, it is a mistake to focus the public's attention on hunting, which seems to strike the imagination more than shooting does, when the suffering inflicted by incompetent and irresponsible shooting is much more severe and prolonged and affects a very great many more animals.

Having determined our policies, we have had to find means to promote them. No useful purpose would be served by trying to duplicate the splendid work done by the R.S.P.C.A. and by kindred societies in Scotland and Ireland ; rather do we give them all the help and support that we can. The niches which UFAW can usefully fill are those which arise out of its special connection with the scientific and educational worlds. This has imposed on our style of propaganda certain restrictions which have proved helpful. David Hume used to go through everything he had written and delete any superlatives which had inadvertently slipped in, and scientists are accustomed to write in the same spirit, avoiding words which are laced with strong feelings.

UFAW's publications have also gained much from the good humour which has been one of their recognized characteristics, and this has been mainly due to the influence of its Honorary Artist, whose personality expressed in his drawings has been a potent factor in moulding UFAW's mentality ever since the year in which that organization was founded. And perhaps the most important rule of all for avoiding quarrels and resentments is this : *never to impute motives*. It is found that ill feeling can be disarmed when the truth is ascertained impartially and is told without spite.

Most people, in this country at all events, have some sympathy for animals ; those who have none are either neurotic or imperfectly educated. With this fact in mind, UFAW tries to enlist the help of persons who are actively engaged in occupations which entail a risk of suffering for animals. Examples are afforded by the co-operation of Hector Whaling, Ltd., in connection with electric whaling ; of the British Field Sports Society in connection with shooting ; of some farmers in connection with rabbit-trapping ; and of experimental biologists in connection with laboratory animals.

An objective and realistic approach to controversial topics has obviated many difficulties. UFAW counts among its members both vivisectors and anti-vivisectionists, Masters of Hounds and opponents of blood sports, vegetarians and carnivores, and yet it has been free from the quarrels to which a fondness for animals often gives rise. This result has been achieved by striving for objectivity both in matters of policy and in matters of fact.

Blind Spots

First published in 1946.

THE WHOLE world was horrified and shocked by the cruelty practised in enemy concentration camps. How was it possible that human nature could descend to such depths? Did people who did such things belong to the same species as ourselves? Would we ourselves condone such atrocities for a single moment in any circumstances? Why did the ordinary German and Japanese citizens fail to rise up in wrath and make these crimes impossible? The answer to these questions is instructive. It lies in the effect of familiarity and of conformity to an authorized moral code in blinding the conscience to abuses that have become customary.

John Newton, confessor of William Wilberforce and author of the hymn beginning "How sweet the name of Jesus sounds", had a most extraordinary history. In his youth he was a worthy prototype of the disreputable Bill the Sailor, famed in military song; in an intermediate period, from 1748 to 1754 and after his religious conversion, he was mate and subsequently master of a slave ship. What happened to the slaves on such ships he described 34 years later in his *Thoughts upon the African Slave Trade*: "Their lodging-rooms below the deck are sometimes more than five feet in height and sometimes less; and this height is divided towards the middle, for the slaves lie in two rows, one above the other, close to each other like books upon a shelf. I have known them so close that the shelf would not easily contain one more. The poor creatures are likewise in irons, for the most part both hands and feet, and two together, which makes it difficult for them to turn or move. They are kept down by the weather to breathe a hot and corrupted air, sometimes for a week." No sanitation was possible and epidemics were frequent. Newton added: "I believe, upon an average between the more healthy and the more sickly voyages, one-fourth of the whole purchase may be allotted to the article of Mortality." While this was going on, the pious master of the ship, who treated both sailors and slaves as humanely as the exigencies of his calling allowed, repressed swearing and profligacy among his crew and read the liturgy with them twice every Sunday. In his *Authentic Narrative*, published six years after his last voyage in command of a slaver, he wrote: "During the time I was engaged in the slave trade I never had the least scruple as to its lawfulness. It is indeed accounted a very genteel employment. However, I was sometimes shocked with an employment that was perpetually conversant with chains, bolts and shackles." Eventually it dawned on him that the slave trade was un-Christian and he became a pioneer, if not the originator, of the movement for abolishing it; but it is instructive to see how at one stage his conscience, though sensitive in many other directions, was blinded by custom in this particular direction. The founders of the trade were worthy men, too. Such was Sir John Hawkins, who by its means opened a way to great wealth for Bristol, Liverpool, London, and indeed the whole realm. His voyages in quest of slaves were made in a ship, the *Jesus*, which was lent to him for the purpose by Queen Elizabeth. He was knighted for his services in 1565 and took a crest which symbolizes their nature. Stow says that in his epitaph in St. Dunstan's in the East, which was destroyed in the Fire of London, he was described as "One fearing God and loyal to his Queen, True to the State by trial ever seen,

Kind to his wives . . .". The fact is that the essential difference between a reputable citizen and a criminal is that any blind spots which the former may have relate to abuses of which contemporary law takes no cognizance.

Many similar instances could be cited. In ancient Rome crucifixion, from which our word "excruciating" is derived, was a normal punishment for slaves and foreigners, and although cultured Romans disliked the practice the possibility of abolishing it never occurred to them. Cicero called it *crudelissimum teterrimumque*, a most cruel and disgusting punishment, but he did nothing about it even when he was consul. Seneca, in the essay *De Clementia* which

he addressed to Nero, opposed a good many cruelties, including even unnecessary cruelty to slaves, but he seemed to take crucifixion for granted; offended by the practice of drowning a parricide in a sack containing a dog, a cock, a serpent and a monkey, he complained that at one time it became so common that sacks were as frequently to be seen as crosses (*De Clementia* I 23-1). Again, our very respectable forbears regarded torture as a normal instrument of legal procedure; for instance, they tortured Guy Fawkes to extract a confession. Their chimneys were swept by orphans known as "climbing boys", who were selected from the poorhouses at the age of about 8 years; undersized children were preferred because many chimneys were only nine inches wide, or sometimes seven. Any claustrophobic hesitation was overcome by beating or by lighting fires under the novices. Sometimes they got stuck in the position they called " nose and knees together ". Sometimes they were suffocated by soot, or burned when the soot was smouldering. They lived in dirt and misery, without schooling or religion, and contracted a painful disease of the scrotum known as " sweep's canker ". Half a century of agitation was needed before housewives would consent to do without such a useful institution ; yet these apparently callous ancestors of ours were decent, humane, civilized people like ourselves. They simply had blind spots towards cruelties that were sanctioned by custom.

All this raises an interesting question. Is it possible that there may still be blind spots in the Anglo-Saxon conscience? Will some practices in which we see no harm be classed by our descendants along with the slave trade and crucifixion? The late Commander Edward Breck, United States Navy, spent the latter part of his life in fighting against a practice which is widespread among the English-speaking peoples and was described by him as " the most awful horror in the history of the world ". He was referring to the use of the steel trap, or gin trap. Whether or not he was exaggerating when he said it was " the most awful horror in the history of the world " must be a matter of opinion in the absence of agreed definitions and of a system of measurement. On one hand the victims are not human beings but animals, and therefore their sufferings are not replicated in the lifelong memories of a surviving parent or child, *cujus animam gementem pertransivit gladius*. But against this advantage must be offset the stupendous scale on which the traffic is carried on. Exact figures are not available, but an estimate can be made from various trade returns and other sources of information. The harvest of furs trapped in the New World, Australasia and Europe totals more than 100 million in a normal year; these are distinct from the skins of domestic animals and silver foxes, which are farmed. As a luxury trade the importation of furs into Britain was repressed during the war, but it has been re-established with official encouragement on account of the profits got by processing and re-export.

The effects of the steel trap or gin are very similar to those of crucifixion. In the case of the largest traps intended for lions or bears the similarity even extends to the piercing of the limbs with spikes, as can be seen in a trapmaker's illustrated catalogue. On the other hand the legs of a trapped animal are usually broken at the beginning of the punishment whereas, as Justus Lipsius has shown (*De Cruce* II 14), the leg-breaking or *crurifragium* was applied by the Romans only as an extra penalty; for instance, as a make-weight when the condemned man or *cruciarius* had for some reason to be killed before he had endured the whole of his punishment.

When animals have their legs crushed by the jaws of the trap they lose self-control. Some of them fight the trap and break their teeth on the steel. Some struggle so violently that they twist off a leg and escape. Some are eaten by wolverines, buzzards and other predators while they are held fast; similarly men were sometimes crucified in the arena and simultaneously exposed to wild beasts. These occurrences involve the trapper in financial loss, but they are often avoided by the use of some device such as the spring pole for suspending the victim by its crushed limb; in the same way a *cruciarius* was suspended by his lacerated hands. This technique seems, however, to be restricted to the New World, where in sparsely inhabited regions the trap line may be a hundred miles or more long and the animals may have to be held in traps for one or more weeks.

As in crucifixion, the time during which life can persist under these conditions varies greatly with circumstances and with the individual constitution and resistance to thirst and pain. St. Andrew is said to have lived more than two days when he was crucified, Victorinus three days head-downwards (like an animal trapped on a spring pole), and Timotheus and Maura as much as nine days. Josephus was once able to rescue three of his friends who had been crucified, and although one recovered the other two died in the hands of the physician (*Life*, 75). Similar variations occur in the case of animals. The writer has found animals apparently dead in a trap after one night, but usually they can survive much longer. A trapped lynx is said to have survived three weeks, though such a case would be hard to prove. Lipsius discusses at length (*De Cruce* II 12) whether death in crucifixion was normally caused by shock or starvation (*languore aut fame*).

Thirst plays a part in the sufferings of trapped animals as it did in crucifixion. It is known from the Talmud (*Sanh.* 43 *a*, according to Edersheim) that there was in Jerusalem a humane society composed of women who provided wine and myrrh to quench thirst and to serve as a primitive anæsthetic, but this was exceptional. Grey Owl found a female beaver trapped but suckling one of her young, and when he chopped off the mutilated paw she thirstily drank the blood from it. Suffering may also be aggravated by the onset of septic conditions.

Now we have seen that in former days the slave trade, more inhumane than Belsen or Buchenwald, was accepted as a matter of course; at one time even Quakers kept slaves. The sweeping of chimneys by climbing boys, sweated labour in cotton mills and coal mines, the war by which China was forced to import opium, the death penalty for theft—all these things in their day were tolerated by kindly and respectable men and women. The torture of witnesses for legal evidence was defended by the great Lord Bacon. The modern world, also, suffers from blind spots.

A woman at Buchenwald had a beautiful lampshade made from the tattooed skin of a human victim, but there is no reason to suppose that in other respects she was wanting in good taste; she simply had a blind spot on the subject of Jews and anti-Nazis, in accordance with the sentiments prevailing in the circles in which she moved. Similarly most British women wear furs that have been got by the infliction of prolonged and severe pain, fear and thirst, in the manner described above. They are eminently civilized women. Some of them are to be seen on the platform at meetings of charitable societies, including animal-protection societies, but they have a blind spot on this particular subject. The present writer has sat in church on a Good Friday behind a row of women devoutly wearing furs that may have cost a hundred days of extreme suffering in the aggregate. This was on a Good Friday.

Even when the public conscience has begun to be aroused by a plea for reform, that plea has always met with obstinate opposition. Apologists for the *status quo* normally try to rationalize their case, first by arguing on humane grounds either that the abuse complained of is in fact a beneficent practice or that reform would be more inhumane; and secondly by making much of the economic sacrifice involved, with the implication that reform would be against the public interest. Thus the Duke of Clarence used the humanitarian plea (*Hansard*, March 1st, 1799) that the slave trade was rescuing negroes from savagery, and the Earl of Albemarle (*Hansard*, May 10th 1844) said that if the Chinese were to be deprived of opium " drunkenness and the crimes attendant upon drunkenness produced by ardent spirits would increase, the one ten and the other a hundredfold". Of a Bill to reduce the working week to 58 hours, Sir J. Graham said (*Hansard*, May 10th, 1844) that it would produce "such general poverty, such physical degradation, such absence of all civilized manners and habits, as must lead to general confusion, almost to primitive barbarism, to desperate crime". Of a Bill to prohibit the use of man-traps and spring guns against poachers, Lord Blandford said (*Hansard*, March 23rd, 1827) that "if this measure were agreed to, armed bands must be marshalled against the midnight invaders of property, and the conflicts of the two parties would produce much more human suffering than had ever resulted, or could result, from the setting of spring guns".

Similarly apologists for fur-trapping have pleaded humaneness. They have asserted that animals are trapped by humane means, although the slightest acquaintance with the trap-makers' catalogues or the trappers' manuals and periodicals will suffice to establish the truth. Again, spokesmen of the trade have claimed that trappers are ridding the world of cruel predators; yet in fact the vast majority of fur-bearers—coney, musquash, beaver, squirrel and the rest—feed on vegetable matter only, and those which are predators kill far more quickly than the trap does.

More sincere has been the defence of abuses on the ground of economic advantage and public policy, for although righteousness exalts a nation it often demands some monetary sacrifice. Yet even this line of defence has been marred by exaggeration. Alderman Newman said of the slave trade (*Hansard*, May 12th, 1779) that "if it were abolished altogether, he was persuaded it would render the City of London one scene of bankruptcy and ruin". On another occasion (*Hansard*, April 19th, 1791) he said that abolition would "ruin the West Indies, destroy our Newfoundland industry, which the slaves in the West Indies supported by consuming that part of the fish which was fit for no other consumption, and consequently, by cutting off the great source of seamen, *annihilate our marine*". Of the proposal to abolish climbing boys, Sir John Yorke said (*Hansard*, February 22nd, 1818) that the use of the sweep's brush would destroy the mortar in the chimneys, cause fires, and so necessitate stripping off the tiles and lead to the leaden gutters being stolen. These exaggerations covered a substratum of truth. Orphans from the poorhouse were a valuable source of cheap labour, and the introduction of Glass's Sweeping Machine and other forms of the now familiar sweep's brush necessitated a structural modification of chimneys containing right-angled bends. Again, the West Indian planters at least were hard hit by the abolition of slavery. Some sacrifice has often to be made if one is to do the right thing.

Similarly it has to be admitted that immense wealth is invested in the fur trade, and indirectly all of us benefit by the profits accruing from the processing and re-export of skins. Then, trappers and trap-makers earn a good living, like the former makers and users of racks, thumb-screws and slave shackles. Against the loss of these profits could be set off the advantage that would accrue to the home textile industry if trapped furs were to be replaced by long-pile fabrics, which can give the desired lightness and warmth and need only the whim of fashion to establish their beauty. Fur-farming also would be encouraged, though how far that is desirable is a matter of opinion. Nevertheless it seems likely that, when allowance has been made for all advantages, a monetary sacrifice would have to be made, at all events initially, if the gin trap were to be abolished in civilized countries upon a decision to do what is morally right and to eschew a practice that no conscience can sanction without insincerity.

On the other hand there is little or no validity in the plea that the gin trap is needed for keeping down animal pests. On the contrary, the stock of fur-bearers, including musk-rats, has to be kept up by conservation laws and close seasons in Canada and the United States. As regards Australia, the Special Commissioner appointed by the New South Wales Government to investigate the rabbit menace wrote in his report of 1928: "The trapper, working for a living, does not exterminate the rabbits. . . . He can be likened to a man working a crop of lucerne—he cuts down the crop to a certain extent, knowing that healthy plants are there and that he may shortly come back and take yet another crop" (pp. 10, 39), and more to the same effect. He recommended methods other than trapping for the control of the pest. The situation in Britain will be discussed at the end of this article.

A supplementary weapon used in the defence of abuses is the personal attack on would-be reformers. James M'Queen, writing in defence of slavery, said that Fowell Buxton merely wished, as a brewer, to eliminate the competition of West Indian rum (*West India Colonies*, 1824, pp. 367, 368). The Act for the Regulation of Chimney Sweepers, which abolished the climbing boys, was promoted by Robert Steven of the Hand-in-Hand Insurance office; and the Earl of Hardwicke alleged in the Lords debate (July 6th, 1840) that "it was a conspiracy on the part of the fire offices to increase the number of fires, and so promote their own interest". Monsieur Bard, rapporteur of the Court of Cassation which was to revise the Dreyfus trial, was appointed on the ground that, being the only bachelor available, he alone would be immune from Press ridicule of a certain kind (*Reinach*, IV 324). Nor has the weapon of personal imputation lain unused in defence of the gin trap ; see, for instance, *Hansard*, Lords, May 28, 1935, column 20.

[In conclusion a word must be said about the gin trap in Britain, where it is principally used for catching rabbits but catches a great many other mammals and birds as well, often inadvertently. (For rats it is used little, if at all, by up-to-date staff employed in pest-control.) The rabbit is a pest and as such can be got rid of by humane methods ; by cyanide fumigation when it lives in woodlands. The Ministries of Food and Agriculture agree that the food situation in this country would be the better for the complete elimination of the wild rabbit, though many country-dwellers would prefer a less drastic policy. The reason why rabbits are trapped is that they have a market value, and trappers no more try to get rid of them than butchers try to get rid of sheep. Rabbits are their stock in trade, and have multiplied exceedingly in counties like Carmarthenshire as a result of the introduction of trapping. The aim of humane thinkers in this country must be, first to kill the trapping industry by means of a thorough campaign of rabbit-clearance, centrally organized with adequate funds and an adequate labour force ; secondly to prohibit the sale, use and manufacture of the gin trap in Britain ; and thirdly to publicize that prohibition as an example to be followed throughout the world.]*

* The final paragraph is now out of date (see p. 45), but has been retained for the sake of completeness.

The Gin Trap: UFAW'S Long Battle

Reprinted from *The UFAW Courier* No. 15, Winter, 1958.

AT THE time when UFAW was founded as ULAWS, in 1926, little thought was being given by animal-lovers to the gin trap. The R.S.P.C.A. was passing through a bad patch (from which it has since happily recovered) and was making no progress in this field.

Although ULAWS at once perceived the importance of the problem, we were unable at that time to see a solution of it; and meanwhile there had been formed a small independent body, the Anti-Steel-Toothed-Trap Committee, which organized an annual Rabbit Week and a petition to Parliament but had no practical knowledge to go upon. In 1928 ULAWS enlisted the help of the University of London Union in giving a little publicity to the subject, but in the absence of a practical policy little progress could be made.

The turning point came in 1929 when Sir William Beach Thomas, well-known for his articles on natural history, addressed a meeting organized by ULAWS at the annual congress of the National Union of Students. Beach Thomas put forward the view that trappers, instead of keeping the numbers of rabbits down, kept them up, somewhat as butchers keep up the numbers of sheep in spite of the fact that they kill them; and he held that wild rabbits were doing much more damage than their carcases and pelts were worth. ULAWS adopted these views but was a voice crying in the wilderness at that time, when wild rabbits were almost universally regarded as a valuable source of food; whereas today, as a result of our propaganda and of experience of the effects of myxomatosis, the soundness of our policy is widely accepted—so much so that New Zealand, in order to discourage the rabbit-trapping industry, has devalued the rabbit by prohibiting exportation!

At first our propaganda met with little response, but one of our leaflets happened to fall into the hands of Colonel W. H. Buckley, Master of the Carmarthen Hounds, who had already reached the same conclusion as a result of experience. He at once became a staunch and invaluable ally. The following year Dr Kirkman joined ULAWS and soon became its Honorary Secretary for Wild Life; he spent many weary nights trying out various methods of catching or killing rabbits. We even persuaded the R.S.P.C.A. to market the Lewis Humane Snare, for which I now offer a sincere apology! On the other hand we were on sound lines in advocating gassing (with motor-car exhaust gas at that date) and the long net.

PUBLICITY

In 1931 we arranged a debate at King's College, London, with Professor Julian Huxley in the Chair, the disputants being a poacher, Mr Collington, who was an expert with the long net, and a Gloucestershire rabbit-trapper, Mr Isaac Gough. Collington, who never used a trap, made a good living by selling the products of his illicit chase to London restaurants, with some of which he had contracts; and a legacy had enabled him to buy a small cold-storage plant in which he kept his poached rabbits and pheasants. This debate, which brought a breath of the countryside into the Strand, achieved a good deal of publicity and helped to call attention to the cruelty of trapping, to which few were then giving any thought. *The Field* opened its correspondence columns to letters describing experiences with traps and nets.

In 1932 that great-hearted woman Frances, Countess of Warwick, a famous Edwardian beauty who had been a friend of King Edward VII and retained her charm into old age, organized on behalf of ULAWS a meeting at the Connaught Hotel and invited a number of persons who were, and others who ought to have been, concerned about trapping. Dr Kirkman gave an address and there was a discussion which was published as the first edition of *Man versus Rabbit*, with Fougasse's well-known illustration on the cover. In the following year the Anti-Steel-Trap Committee dissolved itself and handed over its functions, subscription list and liabilities to ULAWS; but the liabilities, amounting to £150, were generously met by a donation from Commander Cather. ULAWS undertook to promote a Bill and for this purpose appointed Dr Kirkman's brother-in-law, Mr H. W. J. Stone, to be its part-time parliamentary agent. (It may be of interest that shortly afterwards the British Science Guild, of which I was an Hon. Secretary, and the Association of Scientific Workers, in which I was Chairman of the Executive Committee, formed the Parliamentary Science Committee, to which ULAWS contributed Mr Stone; and later on this Committee, on the outbreak of war, was dissolved and re-formed as the present Parliamentary and Scientific Committee.)

THE FIRST PARLIAMENTARY BILL

We duly got out a draft Gin Traps (Prohibition) Bill, and circulated copies by way of propaganda. One of these fell into the hands of Sir William Graham-Harrison, K.C., who was perhaps the finest parliamentary draftsman that this country has ever had, and had been mainly responsible for drafting the constitution of the British Commonwealth. He strongly criticized our draft and offered to make us another without any fee; and from that time onwards his services were always at our disposal in an honorary capacity. His draft was a miracle of lucidity and simplicity, and when later on it was debated in the House of Lords not a single criticism of the wording could be thought of by our opponents.

CYANIDE GASSING

But before the Bill came before Parliament a double event of far-reaching importance for our purpose had occurred. The late John McMath, an Australian student at King's College, London, whose devoted services to UFAW deserve always to be had in remembrance, had acquainted us with the use of cyanide gassing in Australia and had put us in touch with Mr W. T. Stead. Stead had been appointed by the Government of New South Wales to study the rabbit problem, had reached the same conclusions as ourselves as to the effect of the rabbit-trapping industry, and was an ardent advocate of cyanide gassing. He became a staunch ally, and we republished some of his writings in this country. Then in 1935 Colonel Buckley quite independently hit upon the idea of killing rabbits with Cyanogas, sold by George Munro, Ltd., for gassing rats, and tried it out with conspicuous success. He next took the important step of getting Imperial Chemical Industries interested in the method, and they produced Cymag which, like Cyanogas, produces HCN on reacting with the moisture of the air. ULAWS now went all out on cyanide gassing, which to-day is the standard technique for killing rabbits in their burrows.

Our bitterest and most effective opponents at that time were the British Field Sports Society, who were worried about our policy for two reasons. One was that they were afraid that foxes would be gassed, to the detriment of hunting; but Buckley, himself a Master of Hounds, pooh-poohed this argument and has proved to be right. The other reason was that game-keepers relied on the sale of trapped rabbits for a part of their emoluments, disclosed or undisclosed, and accordingly they very emphatically and with many arguments advised their employers against gassing. At a later date their journal, *The Gamekeeper*, gave instructions for restocking land on which rabbits had been destroyed as a pest by the agricultural authorities.

In preparation for the Bill, *Man versus Rabbit* was completely rewritten. It now contained conclusive evidence that the rabbit-trapping industry was responsible for the plague of rabbits. Rabbit-damage had by this time come to be much more generally recognized as an evil, whereas in 1918 the old Board of Agriculture had published instructions for farming wild rabbits. A questionnaire which was circulated by ULAWS contained the question "Are rabbits regarded as a pest in your locality?" and two-thirds of the replies were "yes" while one-third were "no". One of the most damning arguments in the book was Buckley's history of the growth of the rabbit plague in his own county of Carmarthenshire; it left no room for doubt that wherever trapping had been introduced the rabbit population had increased by leaps and bounds.

IN THE HOUSE OF LORDS

Publicity, the questionnaire, and Colonel Buckley had by this time put us into touch with a large number of farmers and others who could speak from personal experience. One of these was David Perkins of Fishguard, who "oozed prosperity" as was said, and had rabbit-free land in the midst of a county that had become one huge warren; and he had achieved this result by banning the gin trap and killing rabbits by other methods. We published a good many leaflets and the like, written by Colonel Buckley and others, and by this time had aroused a vast amount of public interest and support. The R.S.P.C.A. was co-operating fully, as the Scottish S.P.C.A. had always done, and when our Bill came before Parliament we had the support of some ninety societies and public bodies.

One Saturday in the hot spring of 1935 I received a telegram from Buckley: "Fox arriving Paddington Sunday morning; please collect." I found a more-than-dead fox whose pad had been cruelly mangled with a trap, and next came a letter suggesting that I should use it for propaganda. I wrapped it in brown paper and took it on the top of a bus to Trafalgar Square, after warning a press agency. The weather was warm, the fox was fragrant, and the people on the bus looked perplexed. I put the fox on the plinth of the Nelson Column with a trap on its leg, and the pressmen took photographs. We used these for propaganda, but the press publicity was disappointingly small; except that a nature writer in *The Morning Post*, writing about wild life in the London parks, remarked that recently a fox had dragged a trap into Trafalgar Square.

The Gin Traps (Prohibition) Bill came before the House of Lords in May, 1935. It was introduced by the late Viscount Tredegar, a wealthy young peer who was related to most of the hereditary peerage, and it was hoped that he would carry them with him. Unfortunately he was a poet who had mixed a good deal with Italians, and consequently did not draw any sharp distinction between fact and fancy. So long as he stuck to his brief all was well, but when he left it his poetic fancy took over. He said that the R.S.P.C.A. had provided him with humane rabbit-traps which had been completely successful on his estate;

and afterwards when people wrote to him for details he passed the letters on to me for answers. As neither the R.S.P.C.A. nor anybody else knew of any such trap at that time, this was embarrassing. But the debate was exciting. Lord Merthyr made a fine speech; he had kept his own estate in Pembrokeshire free of rabbits, prohibiting trapping on it. A particularly interesting intervention was one by Lord Astor which was quite unexpected and weighed heavily on our side. The opposition came from those who were put up by the B.F.S.S. and coached by their own gamekeepers, notably a distinguished peer who "trembled to think" what would happen if the gin trap were abolished. *The New Statesman* in its next issue advised him to "stop trembling and start thinking". Our opponents sought to side-track the Bill by offering the prohibition of open trapping (*i.e.*, of rabbit-trapping elsewhere than in rabbit holes) and this gesture achieved its fell purpose in the end. Seeing that the battle was going against us, we advised Lord Tredegar to agree that if the Bill should pass a second reading it should go to a Select Committee before being proceeded with, but we lost the Bill by 42 votes for us and 46 against. The attendance was the largest in the whole of the session, larger than for the India Bill. Some of the backwoods peers had never, I believe, attended a debate before.

THE MERSEY SELECT COMMITTEE

The numbers of rabbits continued to increase, with a fall in price, and in the Annual Report for 1935-36 we were able to say "This has helped farmers to realize the truth of one of our contentions—that rabbits as stock do not pay. This element in our position is now a widely held opinion, and the plea that rabbits are a valuable source of food commands far less assent than it did three years ago."

Meanwhile support for our campaign had reached enormous proportions. A meeting of the National Federation of Women's Institutes, 7,000 strong, passed a resolution for abolishing the trap with very few dissentients. The Scottish S.P.C.A. arranged for numerous demonstrations of humane gassing.

Major Van der Byl (whose Fur Crusade was later left to UFAW in his will) is mentioned for the first time in this Annual Report.

In December, 1935, Lord Tredegar called together a few of our supporters in the House of Lords and arranged for Lord Merthyr to introduce a motion, with terms drafted by Graham-Harrison, calling for a Select Committee on Damage by Rabbits. This was accepted by the Government, and the Mersey Committee was accordingly appointed.

It may seem odd that a humane society should have pioneered the view that wild rabbits should be treated solely as a pest. The explanation is that wild rabbits were being farmed in order that every year between thirty and forty million of them might be tortured to death in traps and thus put money into the pockets of the rabbit-trapping industry. This industry prospered because farmers said " the rabbits pay my rent ", and because gamekeepers said " rabbits are a valuable source of food ", and because hatters used some 20 million rabbit skins per annum for making felt hats. If once we could persuade the public that, from a national point of view, rabbits do not pay, all justification for the rabbit-trapping industry would disappear. As this view gained support, however, our opponents changed their ground. They now alleged that trapping was necessary " for keeping down rabbits ", though, as our poacher friend Collington once said, " offering to keep down a farmer's rabbits would be like offering to keep down his sheep ". In practice farmers grumbled about the rabbits in the summer but preserved them in the winter. They wanted it both ways but could not have it so.

The Annual Report for that year (1935-36) mentions myxomatosis for the first time, but expresses doubts of its efficiency and humaneness. It also mentions that in Nazi Germany cruelty to animals, including gin-trapping, was punishable with a fine of £500 and two years' imprisonment.

In April, 1936, *The Times* published a letter in which Sir Rowland Sperling pilloried damage by rabbits and invited co-operation. ULAWS got into touch with him and called, at the Natural History Museum, a meeting of his sympathizers who delegated to ULAWS the task of organizing evidence for the Mersey Committee. The latter sat in December of that year.

The R.S.P.C.A most loyally co-operated, even suspending their humane-trap competition during the hearings in order not to side-track the line of evidence we were putting forward.

Unfortunately, being more accustomed to scientists than politicians, we made the elementary mistake of supposing that the Mersey Committee would be interested in factual evidence. It turned out that they were not in the smallest degree interested in factual evidence. They were interested to see what measure of agreement could be reached among rural personages, and of all the witnesses who appeared before them the only ones who offered factual evidence were Sir Rowland Sperling, those who represented I.C.I., and our own witnesses. Our memorandum of evidence was supported by statements made by 33 farmers and one professional trapper, and these witnesses' statements were severely restricted to their personal experiences. We brought up a selection of farmers from various parts of the country to give oral evidence ; by this time we had the help of a retired marine engineer, Lemuel Parker, with time on his hands, and it was his job to round up and shepherd our witnesses. The night before the hearing we took them to a farce in which Yvonne Arnaud was the much-married heroine juggling with a plurality of ex-husbands, and after the play she received us all on the stage. It was a great night for some of our rural friends who were visiting London for the first time. Our chief witness was Colonel Buckley of Castell Gorfod, supported by David Perkins of Fishguard, and Major Francis, an estate agent in a large way. Unfortunately the members of the Select Committee, not being interested in facts, sought to entangle them in theoretical arguments which somewhat side-tracked the main issue.

The Committee's report formulated a policy based on a compromise between conflicting opinions and interests, and this was useful. But it then sought to justify the policy by a series of reckless statements which had no relation to factual evidence but were based on expressions of opinion and unconfirmed fancy which the more influential witnesses had put forward; and some of these statements were demonstrably untrue. The report may, in fact, be regarded as a good summary of the uncritical prejudices which were then prevailing. Nevertheless, rural opinion had become alive to the seriousness of damage by rabbits, and by giving expression to that aspect the Mersey Committee enabled a step forward to be taken.

RABBIT RESEARCH

Charles Elton, Director of the Bureau of Animal Population at Oxford, used to tell us that what was needed was scientifically-controlled research, and accordingly at the Nottingham meeting of the British Association for the Advancement of Science in 1937 I read a paper entitled "Some Facts and Queries relating to the Wild-rabbit Problem", which summarized the available knowledge and made an appeal for just such research. It was the starting point of the ecological work that has since been done in this field. We afterwards published this paper and it had a very good press. (It included a mathematical appendix which won the " Carrot of the British Ass ", awarded every year by Hannen Swaffer for what he considered to be the most asinine of all the follies with which scientists were mystifying journalists like himself.)

Not long afterwards Charles Elton told us that a young zoologist, H. N. Southern, could study rabbit ecology at the Bureau if we would help by making a quite small annual contribution to the expenses, and this we gladly did. Southern, who is now one of the most distinguished of British ecologists, did admirable work for a year or two, but then the war intervened and the research had to be abandoned for the time being.

THE DAMAGE BY RABBITS ACT

By this time ULAWS had joined the Parliamentary Science Committee, and on 18th November, 1937, it organized a meeting of Members of Parliament for agricultural constituencies, with Sir Joseph Lamb in the Chair. I addressed it on " technical and legislative aspects of the rabbit problem in agriculture ". Keen appreciation of the work of ULAWS in its agricultural aspect was expressed. It was indicated that Members interested in the subject would work in close association with ULAWS, though it could not be assumed that all would accept the whole of the ULAWS policy in relation to the gin trap.

We decided to get a Bill introduced and to base it on the recommendations of the Mersey Committee. Sir William Graham-Harrison made a superb draft. Fougasse and I invited the then head of the Land Branch of the Ministry to lunch and asked for Government help with the Bill.

Whereupon our guest waved aside Graham-Harrison's draft and drew out of his pocket an alternative draft which, as could be seen at a glance, was an excessively bad one. He said, in effect, "take it or leave it". If we promoted it as a Private Member's Bill, the Government would help us. The Cabinet had approved it, and any delay would cause us to miss the boat for that session. We yielded to *force majeure* and Lord Sempill agreed to introduce the Bill. It passed a second reading but was so appallingly drafted that shoal upon shoal of amendments came flooding in; amendments not of principle but of drafting. The situation was so bad that, in agreement with the Ministry, a private and unofficial committee meeting was called at the House of Lords for all those who had submitted amendments, in order to sort out the overlap. It was attended by officials from the Treasury and from the Ministry and, on behalf of UFAW, by Sir William Graham-Harrison, K.C., Mrs Ramsbottom (then our Hon. Parliamentary Secretary), and myself. The meeting began in some confusion, but gradually Graham-Harrison's quiet voice established an ascendency. The principal Treasury draughtsman had been his subordinate and regarded him with something like awe, and in the end he completely dominated the proceedings. The Bill was amended again and again, in Committee, on report, on third reading, in the Commons (where we entrusted it to Sir Joseph Lamb), and on return to the Lords. Then came the war and the Act was superseded by a Defence of the Realm Regulation, but the essential feature has always been preserved; that is, the power conferred on the Ministry of Agriculture to take effective steps for keeping down rabbits.

DURING THE WAR

The Ministry of Agriculture, I regret to say, advocated

that rabbits should be trapped for food in the winter, and then gassed in the summer. On the other hand it wisely gave farmers a 50-per-cent subsidy on the use of Cymag, and with the co-operation of the Scottish S.P.C.A. and the R.S.P.C.A. and of Messrs George Munro we continued to publicize gassing which gradually came into favour. UFAW's *Instructions for Dealing with Rabbits* went through successive editions and was widely used. The staff of UFAW had disintegrated, leaving Dr Vinter almost single-handed, but nevertheless the work continued, including considerable propaganda and a questionnaire addressed to natural-history societies about the way the new powers of rabbit-control were working out. The Annual Report for 1944-45 says "Instructions for re-establishing stocks of wild rabbits were broadcast by the BBC on 15th September, 1945": we had a long way to go yet.

CONTINUATION OF RESEARCH

On 10th May, 1945, Charles Elton invited representatives of UFAW to meet his staff at the Bureau of Animal Population in order to discuss the research needed for the humane control of rodents. (In those days rabbits were still rodents; they have since evolved into lagomorphs.) Southern was now involved in other work, but Worden, at that time Milford Professor of Animal Health at Aberystwyth, offered to put one of his young graduates, Winnie Phillips, on to rabbit research and to supervise her work, UFAW paying her salary. She stayed with us for four-and-a-half years, receiving much help from H. V. Thompson of the Ministry of Agriculture and from the Bureau of Animal Population, and established *inter alia* the facts that (1) rabbits damage the quality as well as the quantity of pasture to a formidable extent, which she measured; and (2) professional rabbit-trapping removes only about one-third of the population, leaving the rest to breed. She also supervised the gassing and ferreting of a farm at Farthing's Hook which was so infested with rabbits that the Pembrokeshire Agricultural Committee had had to take it over. The work was successful and had been closely observed by H. V. Thompson, who next persuaded the Ministry to carry out a pilot scheme of rabbit-clearance in the land around Mathry, in Pembrokeshire. In connection with this scheme he persuaded UFAW to appoint as scientific observer another of Worden's students, Marie Stephens (now Mrs Tom Kind). Any one farmer could spoil the scheme, and some of them were reluctant to come in ; the officials could do nothing with them. But by good luck our two research workers were both unusually beautiful and charming young women who spoke Welsh, and even the most recalcitrant old Welsh farmers surrendered to their cajolery in the end. They also twisted the working party round their little fingers, and set an example of untiring hard work in all weathers ; it made me quite sick to see their fair hands gutting rabbits. It was in large measure to them that the success of the Mathry scheme, which was the parent of all subsequent rabbit-clearance schemes, was due. Both girls received M.Sc. degrees for their researches on the ecology of the wild rabbit.

THE SPRING TRAPS BILL

In 1949 the Home Office appointed a Select Committee on Cruelty to Wild Animals under the Chairmanship of Mr John Scott-Henderson, K.C. UFAW presented a memorandum, and oral evidence was given by Buckley, Worden, and myself, mainly about trapping. Unlike the Mersey Committee this one was interested in facts and the evidence on which they rest, and we had an excellent hearing. The Scott-Henderson report strongly condemned the cruelty of the gin trap, in a way which encouraged us to ask Lord Elton to introduce another Bill. This, the Spring Traps Bill, had its second reading in the House of Lords on 28th November, 1951. By accident or design the Government allotted a date on which many of their supporters were to be in the House for other business, and when the division was called peers came pouring in from all over the building and mostly followed their leaders into the not-content lobby. Having been turned off the floor while the division was in progress, I had to wait outside the door. One peer arrived too late and was locked out. I asked him which way he had meant to vote and he replied "I don't know. What are they voting about?". We lost by 49 votes to 22. Admirable speeches were made by Lord Elton and Lord Merthyr, and we reprinted these for propaganda purposes.

"HUMANE SUBSTITUTES FOR THE GIN TRAP"

Our view that rabbits must be cleared off farms and not exploited commercially had become generally accepted, but friends of the gin trap now alleged, against all the evidence, that traps were necessary for keeping down rabbits. The truth is, however, that, although trapping is the most profitable method of exploiting rabbits for the market, it is inefficient as a method of keeping down their numbers. While you are trapping the cream off a colony of rabbits, the remaining two-thirds are being scared away by the screaming, to come back later on and breed. The yield of trapping then falls off, and in actual rabbit-clearance schemes more efficient though less remunerative methods had had to be adopted when that stage was reached. (The correct methods are detailed in our *Instructions for Dealing with Rabbits.*) Hence the demand for a "humane substitute for the gin trap" becomes irrelevant when the rabbit's true status as a pest is recognized. Again, the difficulty of devising a humane killer trap increases rapidly with the size of the victim. Break-back mouse-traps can be reasonably humane, but the moment of inertia of a striker arm increases as the fifth power of the linear dimensions. There will never be a 100-per-cent kill with any portable rabbit trap. Nevertheless the superstition that a humane substitute for the gin trap was necessary, which had been encouraged by well-meaning humanitarians, had been a main obstacle to reform. It was a prejudice that still persisted, and it had somehow to be bought off before the gin trap could be made illegal. The Government had to bow to this prejudice, but were now genuinely anxious to outlaw the gin trap. They accordingly appointed a committee to develop a humane substitute, and starting from the Sawyer trap, for which the R.S.P.C.A. had awarded a prize, they arranged for its development into the Imbra trap. And although convinced that portable traps of any kind are the wrong means for dealing with rabbits, I must here pay a warm tribute to both the R.S.P.C.A. and the Ministry for this development, without which it would have been impossible to beat down the last-ditch defenders of the gin. The Imbra trap kills about 90 per cent of rabbits outright, and though it maims the remainder it has the great merit that it is ill adapted for commercial trapping, so that, provided the present law can be enforced, rabbit trapping of any sort is likely to die out. There will still, however, be cruel snaring if the rabbit population is not efficiently kept down.

MYXOMATOSIS

In 1953 myxomatosis appeared and, unpleasant though this disease is, it proved to be a blessing in disguise ; for it nearly annihilated the rabbit population, and the resulting bumper harvest at last convinced even the least enlightened farmers that wild rabbits do not pay and that the rabbit-trapping industry had been a poor substitute for good farming.

PROHIBITION OF THE GIN TRAP

In the same year the Government introduced their Pests Bill, Clause 8 of which dealt with trapping. We were much dissatisfied with this Clause, which could have deferred the abolition of the gin trap indefinitely, and we conducted a vigorous propaganda about it. However, it was eventually amended, assurances were given as to the Minister's intentions, and the gin trap became illegal in England and Wales on 1st August, 1958.

Soldiers and Laboratory Animals

An Analogy for Experimental Biologists

Introductory paper read at the UFAW symposium on Humane Technique in the Laboratory and printed in *The Lancet*, February 22nd, 1958.

AT THE time of the battle of Paschendaele there was current a story about a general who earned for himself the nickname of " Plenty-more-men ". It was said that on one occasion a staff officer remonstrated against the reckless way in which he was squandering his reserves, but got the reply " Oh, there are plenty more men, plenty more men ". Whether this story is true or not, we can all think of some biologist or other who deserves the nickname " Plenty-more-animals " ; if not in our own country, then in some foreign country where funds for research are available *en veux-tu en voilà.*

The analogy between a military commander who has to send his troops into battle and an experimental biologist who has, let us say, to ascertain an LD50* by some appropriate procedure is close enough to deserve consideration.

* * * * *

From the point of view of his superior officer, a soldier has two entirely different aspects, which I shall call respectively his " cold-blooded " and his " warm-blooded " aspects, and it is his superior's duty to take account of both. In his cold-blooded aspect the soldier is a regimental number on the strength of a certain battalion. When his battalion is sent into action he automatically goes with it. The expected percentage of casualties will have been estimated in preparing the plan of battle ; in other words, his chance of becoming a casualty is a cold arithmetical factor in a

* LD50 is the dose of a toxic substance which will kill 50 per cent of the animals tested.

statistical calculation. If he does become a casualty he is a detail to be deducted from the ration strength of his unit and accounted for in the parade state. That is his cold-blooded aspect.

A laboratory animal also has its cold-blooded aspect. In this aspect it is just one of a sample of, say, ten, duly numbered in the records of the animal-house; it is one point on a regression line, one unit in a quantal response. In evaluating the results of the experiment such details as its litter number, age, sex, genetic history, and physical response to a given procedure are all that are usually taken into account. The fact that it has a personality is usually overlooked in this connection; the recorded data concerning

it are similar in principle to those which are requisite in relation to chemical samples or reagents, or to the characteristics of thermionic valves in an experimental apparatus.

So much for the cold-blooded aspects of soldiers and laboratory animals. But each of these has a second aspect, which I have called the "warm-blooded" aspect.

Let us take the soldier first. It is a regimental officer's duty to know his men and to look after their welfare. He must inspect their billets and concern himself with their comfort. He must keep an eye on their rations and make a howl if these are unsatisfactory. He must arrange recreation for them. He must even attend, if necessary, to their private affairs, and do what he can for a man whose child is dangerously ill or whose wife has eloped with a stranger. This is the soldier's warm-blooded aspect, and an officer who neglects it is being not only inhuman but also technically incompetent, because morale is of great technical importance and nothing is more destructive of morale than indifference to the fact that the men have human feelings as well as regimental numbers. A divisional general cannot, it is true, know all the men under his command, but even he will try to see as much of them as he can and to understand their human problems.

In the same way laboratory animals have a warm-blooded aspect; even cold-blooded animals like xenopus* have one, metaphorically speaking. A competent biologist will pay great attention to the housing, feeding, and health of his stock, and he will manage them in such a way as to avoid exposing them to any mental stress which is avoidable; he must keep them happy as well as healthy. Failure to do this is inhumane, but it is also inefficient, because healthy and contented stock are an asset in any research, and the biologist who was indifferent to the feelings of his animals would to that extent be technically incompetent as well as callous. The analogy can be pressed even further. In the same way that comradely relations between higher and lower ranks in an army are essential to good morale, so a friendly relation between laboratory animals and laboratory staff keeps the animals tame and so makes them easy to handle and free from stress, besides rendering their short lives happy.

A good biologist will, then, pay much attention to this warm-blooded aspect of the animals he uses, first because as a civilized being he will wish to behave in a humane way; and secondly, though only secondly, because it pays him to do so. Humane management of laboratory animals

* A species of frog used in the laboratory.

makes for efficiency in research.

* * * * *

Leaving the animal-house, let us now turn to the laboratory. Many years ago the late Prof. E. S. Starling expressed to me the opinion that a good deal of the physiological research which was being done in a certain foreign country was being vitiated because the physiologists concerned did not trouble to eliminate pain, and so the secondary effects of pain were being confounded with effects of other conditions which these workers were trying to study.

Dr W. M. S. Russell has conveniently classified under three headings, which he calls the "three Rs", the considerations which bear upon the warm-blooded aspect of

animals in the laboratory. These headings are replacement, reduction, and refinement; and we can apply our military analogy to them.

Here is an example of replacement:

In the early days of the first world war every infantryman carried a pair of pliers with which to cut the enemy wire by hand, and in carrying out this operation the infantry were mown down by machine-gun fire. But later on, inanimate high-explosive shells were substituted for these living wire-cutters, and wire-cutting by the guns was a normal phase in every attack. The replacement section of the UFAW symposium deals with the replacement of living animals by non-sentient materials such as micro-organisms, tissue cultures, and even electrical models in some cases.

For Russell's second R, reduction, a military analogy can be found in the clearing of minefields.

A man who served as a liaison officer with the Russians assured me that he had seen a platoon of infantry marched across a suspected minefield for the purpose of making it safe. If this was regarded as an orthodox method the number of personnel required must have been high and the casualty-rate must have been heavy. In any event, the use of mine-detectors makes possible a great reduction in the number of personnel.

Ways of reducing the number of animals required for a given degree of precision include the methodological design of experiments and the control of variance by means of applied genetics and by paying attention to specific environmental factors.

Dr Russell's third R is refinement of techniques for the purpose of reducing discomfort in cases in which discomfort cannot be wholly avoided.

This refinement of experimental techniques finds a military analogy partly in the steps for preventing casualties (the provision of steel helmets and respirators, for instance), and partly in the field medical services—stretcher-bearers, regimental aid posts, field ambulances, casualty clearing stations, and military hospitals. These are concerned not only with the technical advantage of preventing casualties and of getting lightly-wounded men back into the line again; they are also concerned with the humanitarian purpose of easing the sufferings of men who are so severely wounded that they will never fight again. This care for the severely wounded and the dying, though incidentally it has the technical advantage of improving morale, is primarily a humanitarian activity. In the laboratory there may be cases in which no technical advantage is got by obviating pain in an experiment, no error introduced by allowing a rat or a rabbit to recover prematurely from an anaesthetic; but nevertheless the moral obligation to conduct research with a minimum of pain, fear, and stress is a binding one.

* * * * *

Believing this military analogy to be genuinely relevant, I will add one final point. No military commander is entitled to subject his men to risks and sufferings which he himself would not be prepared to endure if, *mutatis mutandis*, circumstances required him to do so. I say "*mutatis mutandis*", and in particular let us, in the laboratory analogue, leave aside the risk of death, because this is irrelevant when one is going to compare animals with human beings; let us consider only exposure to pain, discomfort, or stress. We may bear in mind also that circumstances do not ordinarily require a divisional commander to expose himself to the same risks as the front-line troops, however often he may have done so in his younger days; his duty is now different from theirs. But in principle he ought to be ready to face anything he calls upon them to face. If we turn now to the biologist, we may say that only in rather exceptional cases can he usefully subject himself to experimental procedures in the laboratory, though a number of research-workers have done this from time to time. On the other hand, he may usefully ask himself whether, if he were this or that experimental animal, he would be willing to make voluntarily the sacrifice which he is calling on the animal to make, for the sake of the purpose for which the experiment is to be performed; apart always from any question of life and death, and account being taken only of any pain, discomfort, or stress that may be entailed.

The Nature Conservancy: The Pioneer Work of UFAW

Reprinted from *The UFAW Courier* No. 4, August, 1950.

ON 11 FEBRUARY, 1949, Mr Herbert Morrison announced that a Nature Conservancy was being constituted for the whole of Great Britain, with a committee for Scotland, and would be responsible to the Privy Council. The Chairman was Sir Arthur Tansley, F.R.S., the Director-General was Capt. Cyril Diver, and the sixteen members included other distinguished ecologists, among them Dr Charles Elton, Director of the Bureau of Animal Population.

The initiative was taken by UFAW (then ULAWS) in the years preceding the outbreak of war. It had long been felt that wild animals suffer severely from the haphazard conflict of interests between farmers, sportsmen, trappers and others, and that something on the lines of the Biological Survey (now the Wild Life Service) in the U.S.A. was needed to establish some sort of policy. In 1932 replies to a ULAWS questionnaire had shown that very little was known as to the distribution and treatment of badgers, for instance. The first written reference to the present subject seems to have been in a letter from Dr Charles Elton to ULAWS on 23 December, 1936: "I am thinking of your ideas about a biological survey bureau. In my opinion it should be a fairly practical Government arm, dealing with wild life control and wild life conservation, with an eye also on the control of introduced animals. It would obviously have to consider game management as well, in connection with agriculture." A memorandum was at once got out by ULAWS proposing a Government "Department of Fauna and Flora". Its function would be to advise the Government "with regard to legislation relating to all plants and animals except as concerns their exploitation for profit, to maintain an intelligence service in collaboration with the Bureau of Animal Population as to (*a*) the population-level at which a species ceases to be desirable and becomes harmful and (*b*) the current abundance of wild species having economic importance; to co-ordinate voluntary effort in the formation of National Parks; to study and encourage humane methods of vermin-control." Dr Elton's comment, dated 13th February, 1937, was that "there is certain to be a strong undercurrent of conflict between your society and . . . [certain vested interests]. We do not propose to take any part in conflicts of this nature . . . I am sure the general principle of the scheme is sound and that you will be doing a very valuable job in promoting it." It may be suspected that his great influence as a pioneer in animal ecology did much in private to further the developments which have since taken place. He felt, however, that humane vermin-control was likely to be controversial, and that the subject was rather one which could best be handled by ULAWS.

To launch the idea, a public discussion on "Man's Relation to Nature and his Response" was organized by ULAWS at the University of London on 22nd March, 1938. The handbill read: "The regulation of economically or aesthetically beneficial or harmful species, field sports, epizootics, importation, and commercial exploitation will be examined in conjunction, and with reference to humane requirements." Fifty-two relevant organizations and numerous individuals were circularized. Sir Peter Chalmers

Mitchell, F.R.S., took the Chair; the principal speakers were the Duke of Bedford and Professor Crew, F.R.S. Sir Roy Robinson (Forestry Commission), Mr W. E. Hiley (Royal English Forestry Society), Mr R. M. Lockley, and Alderman C. H. Gardner took part, and ULAWS undertook to act for the time being as a " clearing-house and co-ordinating body for the moment for all those who are interested in the problem ".

A draft scheme for a " National Fauna Office " was got out, and although the war supervened, thought was given to the subject at odd moments.

In September, 1943, UFAW circulated widely a scheme for a " Wild-Life Authority for Great Britain ", but in correspondence it was recognized that a separate authority would be needed for North Britain because " the fauna of

the wilder parts of Scotland includes a particularly fierce species of retired colonel who would probably react by counter-suggestion to any information or advice emanating from more temperate climes ". In January, 1944, a revised version was circulated by the Parliamentary and Scientific Committee to all its members, and the Committee was addressed by a representative of UFAW on 4th April, 1944; it requested UFAW to ascertain what general support was available and report back, which UFAW did late in 1946.

In this memorandum the conflicting interests were classified as (1) Economic and commercial—agriculture, forestry, fisheries, rabbit-trapping; (2) amenities—national parks, rural amenities, field sports; (3) humanitarian and cultural—zoology, nature study, and humane requirements in pest-control. A central co-ordinating authority should make ecological knowledge available, promote the conservation of desired species and, on a quantitative basis, the regulation of harmful species and *develop and bring into use humane methods of fauna-control*. The authority should at first be advisory but might become executive later; should be independent of sectional interests and responsible to the Privy Council, have a scientific staff, and be specifically charged with the promotion of humane methods of fauna-regulation; and might be combined with a nature-reserves authority. Possible methods of action were outlined, and nineteen existing Acts of Parliament relating to wild animals were listed.

Meanwhile, very similar proposals had been put forward by the British Ecological Society, and a conference took place at the House of Commons on 17th April, 1944, between representatives of the two organizations. There was general agreement, and it was noted that six fields of interest must be taken into account: scientific, educational, economic, sporting, aesthetic and humanitarian. UFAW stipulated that "humanitarian" should imply research on humane pest-control; with this understanding the lead was now to pass to the Ecological Society. The latter drew up a memorandum and invited the Chairman of UFAW to address its annual conference in January, 1945: and late in 1946 memoranda by both societies were circulated to the Parliamentary and Scientific Committee. The movement now passed out of UFAW's hands, but the eventual result is as satisfactory as could have been hoped for, in view of the widespread prejudice against animal-protection societies which even UFAW had to contend with. The demand that the Nature Conservancy (as it has eventually been named) should occupy itself with humane methods of pest-control was gradually squeezed out, but meanwhile research on this subject had been begun by UFAW itself on the basis of a conference between representatives of UFAW and the staff of the Bureau of Animal Population which was arranged by the kindness of Dr Charles Elton at Oxford in May, 1945. It is a matter of great satisfaction that sound scientific ecologists predominate among the members of the Nature Conservancy ; one of UFAW's main aims was to bring about a scientific approach to these matters, in which truth has been too much at the mercy of prejudice. Meanwhile it is for UFAW to concentrate on the specific purpose of devising and promoting humane methods of controlling wild creatures, and in this it can count upon, and has received in generous measure, at least the unofficial sympathy and help of ecologists.

The Vivisection Controversy in Britain

This paper was presented at the Tenth Annual Meeting of the Animal Care Panel on 29th October, 1959, at Washington, D.C. (U.S.A.), and, with the kind permission of the Panel, was printed in *The UFAW Courier* No. 17, Autumn, 1960.

FIFTY YEARS ago feeling ran high in Britain over vivisection. In December, 1902, Professor Bayliss was accused by Miss Lind-af-Hageby of operating on a brown dog while it was still conscious, and he resented this imputation so much that he brought an action for libel against her. The court found that the allegation was untrue and awarded very heavy damages, but the anti-vivisectionists did not leave the matter there; four years later they put up in Battersea Park a statue of the brown dog in question with a rather provocative inscription. The statue was connected to a watchman's hut by an electric alarm wire, and when a medical student began to attack it at night with a sledge-hammer the police quickly intervened and arrested ten medical students. Next day a magistrate fined them £5 each, whereupon hundreds of students paraded the streets of London with an effigy of the magistrate, which they first tried to burn and then drowned in the Thames. During the next few days there were bonfires and processions and more arrests of medical students. A correspondent wrote in the *British Medical Journal* that "when a peace-loving student peacefully defaces" the statue "with a hammer, he is doing . . . his strict legal duty to his country and his king". One day after a month of rioting a crowd kept 200 constables busy in Trafalgar Square from the afternoon until 2 a.m., and meanwhile two constables had to guard the statue of the brown dog day and night. The Home Secretary told the House of Commons that the extra police needed to date had represented one day's service of 27 inspectors, 55 sergeants and 1,903 constables. Eventually the Battersea Borough Council decided that they had had enough; the Borough Surveyor, guarded by fifty constables and three inspectors, removed the statue in the dead of night, broke it up, and hid the pieces in his bicycle shed. Nevertheless feeling continued to run very high for a number of years.

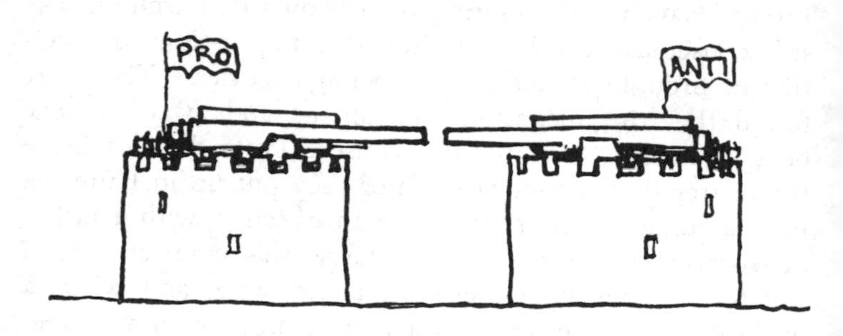

The anti-vivisection propaganda of those days had two bad consequences. (1) In order to win popular support it concentrated attention on cats and dogs, whereas the animals most in need of protection are the rodents, since they are not protected by popularity and are much the most numerous; and (2) it caused tempers to run so high that nobody could have any calm and reasonable discussion with anybody else. Medical scientists developed a sort of persecution complex that made them reach for a gun whenever the subject was mentioned.

But during the past ten years a great change has taken place. Reckless and undiscriminating allegations against scientists have become much more rare and are deprecated by leading anti-vivisectionists, and on the other side of the fence vivisectors can now discuss without passion the ways in which their work could be made increasingly humane. How has this change come about?

At the price of being boastful, I will quote a letter received earlier this year from the head of a cancer research laboratory, who wrote: "I believe, with you, that the climate is changing in this country with respect to rabid anti-vivisection, and this is not only due to the work of Lane-Petter and the Research Defence Society but is in large measure a result of your own enlightened attitude. More strength to your elbow and, through you, to UFAW."

I want to describe UFAW's part in this change, but first let me say a word about Lane-Petter. When he became Honorary Secretary of the Research Defence Society he cut out all the mud-slinging and swashbuckling, and under his guidance that Society has devoted its energies to explaining to the public what medical research is and how it is done in Britain. Moreover, nobody has done more for the welfare of laboratory animals than Lane-Petter has, and thus he has shown that a research-defender can have a heart. In fact he is defending research by the only honest method, namely the removal of legitimate grounds for objecting to it.

In order to explain UFAW'S contribution I must say something about the historic background. In Britain we have a system for putting a check on cruel experiments, and this was pioneered by Charles Darwin with the support of Thomas Huxley, Burdon Sanderson and other eminent biologists of their time. Darwin said that the thought of painful experiments made him feel sick and kept him awake at night, and under his influence the British Association for the Advancement of Science appointed a committee to study the subject. In 1871 this committee made a report which may be briefly summarized thus: no avoidable pain ought to be inflicted; experiments entailing suffering should not be used for teaching purposes; and, in order that no suffering shall be wasted, experiments should be performed only by skilled experimenters in properly equipped laboratories, and painful experiments should not be repeated when once their results have been validated. The same year the British Medical Association adopted a similar report, and a petition signed by leading biologists and calling for legislation was presented to the Government. I should like to emphasize this fact—that the law which we now have in Britain for the purpose of preventing cruelty in the laboratory was a sequel to a petition signed by Charles Darwin, Thomas Huxley, Jenner, Owen, the President of the Royal College of Physicians, the President of the Royal College of Surgeons, and various other leaders of science and medicine. It was based on the recommendations of a Royal Commission which included among others Thomas Huxley, then President of the Royal Society, and Professor Erichsen the surgeon. This law, which we still have, is badly drafted and difficult to understand, but it left a wide discretion to the Home Secretary and, with scientific guidance, that discretion was used to build up a humane and judicious tradition among laboratory workers. It has done a splendid job in the past, though the time has now come when the whole system needs to be modernized in order that it may continue to retain the confidence of biologists and at the same time give more assured effect to the three essential principles on which it is based and which are absolutely indispensable in any genuine system of control. These three principles are, first the licensing of experimenters, secondly the Home Office inspectorate, and thirdly the Pain Rule.

As regards the licensing of experimenters, there has been an effective acceptance of the British Association's recommendation that biological research ought to be done only by qualified scientists. Research consists of one part of experimentation with nine parts of hard thinking, and we do not believe that if an amateur dabbler does enough experiments and uses enough animals he will promote the advance of science, though naturally a research team will include men of various levels of ability working under expert supervision.

Licensing means, among other things, that we do not allow children to do experiments on living animals. On the other hand many British schools do indeed keep pet hamsters, rats and the like, partly in order to teach pupils about the normal lives of the animals and partly in order to inculcate a sense of responsibility for the welfare of our fellow-creatures and to broaden the pupils' sympathies. UFAW has in fact published the standard textbook on this subject with the title *Animals in Schools*, by the veterinary surgeon John Volrath. Again, suitable pupils get a thorough grounding in anatomy, and this entails dissection of dead animals. But experiments which interfere with the normal health or comfort of living animals are absolutely barred out at the school stage. I recently had occasion to sound a number of our most eminent educators and medical scientists on this point, and their opinions were unanimous and emphatic. They were all firmly opposed to any experimentation on animals in schools, partly on scientific grounds but mainly on ethical grounds. Our feeling is that character is more important than knowledge, and since many children pass through a sadistic phase a teacher must avoid the risk of pandering to any morbid interest of that kind, especially at a time when juvenile crimes of violence are increasing at a rapid rate.

The second principle in the British system is inspection by the Home Office. The need for this can best be understood from analogy with clinical research on human subjects. A clinical researcher can become so absorbed in his project that, like Higgins in Bernard Shaw's *Pygmalion*, he will ride rough-shod over the feelings of his subjects.

In a recent symposium on "Quantitative Methods in Human Pharmacology" Dr Fox,* Editor of the *Lancet*, drew a distinction between the "physician-experimenter" whose chief concern is to solve a problem and whose vision "is liable to become so concentrated and limited that he cannot suitably take decisions on the patient's behalf" and the "physician-friend", preferably the patient's family doctor, whose consent should be obtained if the patient cannot understand the proposed experimental procedure. If such an umpire is needed in experiments on human subjects, *a fortiori* he is needed in experiments on animals. To act as umpire in this way is the function of the

* Now Sir Theodore Fox.

Home Office inspectorate with their Advisory Committee consisting of experienced scientists. When a research worker has to decide whether some procedure is justifiable in view of any suffering it may entail, he is almost inevitably biased in favour of his own research; no man may be a judge in his own cause. He needs the Inspectorate to guide him on the basis of established standards of humaneness.

Inspection is a heavy full-time job and the inspectors need not only tact and incorruptibility but also exceptional acuteness, wide experience, great firmness and a strong sense of responsibility. They must enjoy the confidence of responsible scientists. And on the other hand they are there to see fair play for animals, not to defend the Government against ill-informed public criticism. If they were influenced by political considerations of that kind they would be betraying the trust which is imposed on them.

The third essential feature of the British system is the Pain Rule. This was recommended by the second Royal Commission on vivisection and sets a limit to the amount of suffering which may be caused in any circumstances. In spite of all difficulties of definition and enforcement, the Pain Rule does have a practical effect, and various cruel procedures which are practised in foreign laboratories are not practised in Britain. If I were to be asked: "But does not the Pain Rule hamper research?" I would reply, first that even if it did hamper research that would be no justification for abandoning it. The pursuit of knowledge is not the highest good nor does it stand above the moral law. As the Second Royal Commission pointed out, the practice of torturing prisoners to obtain information was abandoned not because it did not yield useful knowledge but because it was immoral. Again, the lethal experiments carried out by Nazi scientists on Jews and political prisoners may have been scientifically valuable but they were morally damnable.

But has British research in fact suffered loss through this restriction on cruelty? If I mention that no fewer than ten Nobel Prizes for Physiology and Medicine have been awarded to British scientists working under the Pain Rule I am not suggesting that the number of Nobel Prizes is a numerical measure of a nation's scientific output; but what I will point out is that a willing compliance with the Pain Rule, in accordance with British law, did not prevent Gowland Hopkins from becoming the father of biochemistry; it did not prevent Sherrington from carrying out his fundamental work on the integrative action of the nervous system; it did not prevent Adrian from elucidating the physical basis of perception; and it did not prevent D. G. Melrose from devising his heart-lung machine for use in open intra-cardiac surgery, to demonstrate which a team of British surgeons were recently invited to Russia. It may even conceivably be the case that having to forgo some experimental techniques has benefited the quality of British science; there used to be a saying in the Cavendish laboratory—"We haven't much money, so we have to use our brains instead." But all this is really a digression; the Pain-Rule restriction must stand, not because it pays to do what is right and abjure what is evil but because this is a duty.

It is against this background that the work of UFAW has been done, and that we have been able to convince our fellow animal-protectionists that scientists can have a heart for animals after all. I come now to the way in which this has been accomplished.

Towards the end of the second world war Professor Worden, a biochemist and veterinary surgeon who had specialized in research on nutrition and who, when a student, had been Chairman of our Students' Council, volunteered to edit a *UFAW Handbook on the Care and*

Management of Laboratory Animals. This achieved immediate success. At first the anti-vivisectionists viewed it with suspicion, but as time went on their attitude changed and they came to see that, since experimentation will continue whether they like it or not, there is everything to be said for making the animals as comfortable as possible. On the other side of the fence research workers were quick to see that this *Handbook* would help them to fulfil their own wishes, by improving both the humaneness and the technical efficiency of their laboratory work. This disarmed the suspicion with which many scientists had regarded UFAW, though some—Julian Huxley and E. H. Starling and the Editors of *Nature* and *The Lancet*, for instance—had stood by us from the very start. This change of climate began to make possible what we had always hoped for—a calm and objective discussion of the problem of vivisection, as opposed to the cut and thrust of controversy in which each side has to make out as striking a case as it can, and nobody dare concede a point or yield an inch of ground.

When a second edition of *The UFAW Handbook* fell due we sought the collaboration of Dr Lane-Petter, who is Director of the Laboratory Animals Centre of the Medical Research Council and also Honorary Secretary of the Research Defence Society. The work of our two joint editors has won world-wide acceptance and, I am sure, is bringing much benefit to the animals.

I shall not describe the contents of *The UFAW Handbook*, because in this assembly I am an amateur addressing experts on the subject with which it deals. I only mention its publication as having been a powerful factor in creating an atmosphere in which it is now possible to discuss vivisection in Britain without provoking a breach of the peace. We also distribute two excellent publications of the Animal Welfare Institute of New York, namely, *Basic Care of*

Experimental Animals and *Comfortable Quarters for Laboratory Animals.*

Some half-dozen years ago UFAW passed on from animal husbandry to the far more controversial topic of experimental techniques, and in the studies which resulted we have enjoyed the sympathy of both vivisectors and anti-vivisectionists.

In this task we looked for inspiration to the great pioneers of modern British biology. As I have already mentioned, in 1870 the British Association, on the initiative of Charles Darwin, laid down certain rules, one of which was, in effect, that there must be no wasteful use of animals, and this view was endorsed by the British Medical Association. Thomas Huxley and the other members of the first Royal Commission unanimously condemned what they called "purposeless cruelty", by which they meant not only "experiments attended with great pain" but any use of experimental animals which was wasteful, inefficient, or devoid of adequate scientific justification.

Thus the founders of modern biology in Britain set their faces against the notion that in science quantity may take the place of quality; that large numbers of experiments with large numbers of animals may take the place of hard thinking in combination with a modicum of judicious experimentation. I do not forget that chance observations have occasionally led to new lines of thought in science but, as Pasteur said in his inaugural address at Lille: "In the field of observation, chance only favours those who are prepared."

More recently, Piaget has found that in solving a simple mechanical problem children under twelve tend to use trial-and-error methods, while children over twelve tend to use their brains. This fact suggests that the pseudo-scientist who pins his faith to reckless and wasteful experimentation has a pre-adolescent mentality.

On the whole, British biology has tended to accept this humane tradition of its pioneers. On the other hand it now uses possibly half a million rats and some two million mice every year, and such large numbers entail an obvious risk that an experimenter may begin to think of animals as if they were things, as if each individual rodent, with its own feelings and its own individuality, were of no more consequence than a test-tube. The possibility that animals may be used in unnecessarily large numbers seems, therefore, to call for special consideration, but various other aspects of experimental practice also need to be studied systematically by humane scientists. Five years ago we accordingly appointed Dr W. M. S. Russell, a brilliant young zoologist who happens to be also a psychologist and a classical scholar, to inaugurate a systematic study of laboratory techniques in their ethical aspect. The first overt result was a symposium on Humane Technique in the Laboratory held in May, 1957. About 120 vivisectors attended, and the Chair was taken by Prof. P. B. Medawar, F.R.S. The papers were published by the Laboratory Animals Centre of the Medical Research Council, and I would be glad to present any member of this Panel with a copy.

A more recent event has been the publication of a remarkable book by Russell and Burch entitled *The Principles of Humane Experimental Technique*. This deserves to become a classic for all time, and we have great hopes that it will inaugurate a new field of systematic study. We hope that others will follow up the lead it has given, and that a generalized study of humane technique, as a systematic component of the methodology of research, will come to be considered essential to the training of a biologist in the way in which spectrometry and differential equations are to that of a physicist, or in which the periodic table and the phase rule are to that of a chemist.

Russell has conveniently classified the problems of humane technique under what we call the " three Rs ", The first R is replacement of conscious animals by unconscious material. Two outstanding examples, discussed at length in the book, are the use of tissue cultures (which we owe to an American scientist, Ross Harrison) and the use of micro-organisms, but on the theoretical side Russell has a good deal to say about what he calls the " high-fidelity fallacy ". It is often assumed that, because a given mammal resembles man in many ways, it will also provide the highest discrimination for some property we wish to test. But this is not generally true. For example many anti-tumour compounds which are effective in laboratory animals have no effect on the majority of human neoplasms, and the antibiotic cycloserine, which is active against tuberculosis in man and *in vitro*, is inactive in guinea-pigs and mice. Humane biologists who search for non-sentient materials to replace live animals may be rewarded by finding subjects with higher discrimination for some particular property.

The second R is *reduction* ; that is, reduction of the number of animals required for a given degree of precision, and this can be accomplished in at least three ways. First there is the economy in sampling which can be effected by the statistical design of experiments and is associated particularly with the names of Sir Ronald Fisher in England and Abraham Wald in America. Elsewhere I have called attention to one famous experiment which used five times as many mice as were needed to give a highly significant result, so that four-fifths of the mice were wasted. Statistical theory is still insufficiently understood by some biologists, perhaps because insufficient importance was formerly attached to mathematics in their training.

Another well-known method of reducing the requisite

number of animals is to use genetically homogeneous stocks, and preferably Fl crosses between pure lines.

A third and much less familiar factor affecting sample-size is the psychosomatic influence of specific components in the environment. Since Michael Chance found that the variance in a gonadotrophin assay was drastically influenced by the manner in which his rats were caged, UFAW has been sponsoring continued research under his direction into effects of this type, in the hope that there may be other specific causes of variance which remain to be identified and eliminated.

I now come to Russell's third R, *refinement* : that is, the modification or selection of laboratory techniques so as to reduce to a minimum any pain or discomfort that may be entailed. First there is what he calls " contingent inhumanity " ; distress, that is, which can be avoided if enough skill and care are exercised. Thus, fear can be avoided by handling the animals frequently and so getting them tame. Again, a blunt or barbed hypodermic needle can cause considerable pain, especially in a small animal, whereas a sharp one skilfully used need not do so. Again, to pith a frog humanely requires skill and knowledge, and plenty of preliminary practice on dead animals. Contingent inhumanity can be avoided only if the professions of laboratory technician and animal technician have their proper status, courses of training, and tests of competence.

Next there is what Russell calls " direct inhumanity ", that is, the distress unavoidably arising from certain procedures. This can be reduced and in some cases abolished as a result of systematic study. The introduction of anaesthesia was the greatest step forward in this field ; but if anaesthesia is to be effective it must be administered with skill, knowledge and diligence. UFAW is sponsoring a research on the neurology of anaesthesia which is being carried out by Dr Phyllis Croft at the Royal Veterinary College mainly by electro-encephalography. She has, for instance, been concerned with the risk of unintentional cruelty entailed in the use of muscle-relaxants, and she has written, for publication by UFAW, *An Introduction to the Anaesthesia of Laboratory Animals.*

Toxicity-testing is one of the main fields in which direct inhumanity arises. UFAW is sponsoring in the Department of Pharmacology in Birmingham the trying-out of a technique devised by Dr Michael Chance and depending on the use of fully-anaesthetized mice. We must not count our chickens before tney are hatched, but otherwise I would express a hope that, as a result, very large numbers of mice may one day be spared from a distressing death in the laboratory. Another hot spot for direct inhumanity is the testing of analgesics, and one of our Swiss members, Dr Radouco-Thomas, has devised and validated a test in which the intensity of the discomfort inflicted is kept to a minimum. Again, the atrocities practised in some researches on stress are abjured and detested by British scientists.

For a full preliminary study of humane technique in the laboratory I must refer to Russell and Burch's very remarkable book. I have only sketched the salient points of our policy very summarily for the sole purpose of explaining how it is that UFAW has helped to create in Britain an atmosphere in which vivisection can be discussed calmly. Scientists are willingly co-operating in the study of humane techniques and, seeing that that is so, the leading anti-vivisectionists, while not abandoning their ultimate principles, gladly welcome what is being done and increasingly try to adopt a fair-minded and discriminating approach to the controversy.

Electrocution: A Historical Retrospect

Reprinted from *The UFAW Courier* No. 11, Summer, 1955.

EARLY IN 1926 Mr Paddison, Hon. Humane Slaughtering Adviser to the R.S.P.C.A., sent me a newspaper cutting about one of those rare accidents in which a man who has had an electric shock is thought by bystanders to be dead but in fact is able to hear all they say. What worried Mr Paddison was that unwanted dogs and cats were being electrocuted at a clinic at Islington for which the R.S.P.C.A. was responsible, and he had begun to wonder whether this process was really as humane as it looked. After discussion he tried to persuade his Council to have a scientific investigation made, but he encountered opposition. He therefore asked me to put a question at the Annual Meetings of the R.S.P.C.A. in 1926 and 1927, which I did. In the end he got his way, and I suggested Prof. F. Lloyd Hopwood, Professor of Physics at Bart's, as a suitable investigator.

THE FIRST INVESTIGATIONS

Hopwood found that in the electrocution cabinet, in which current passed through the body from the neck downwards but did not pass through the brain, the voltage was 1,925 V. on open circuit but fell below 200 V. when current was passing. The current was 155-160 mA., but part of it may have leaked through imperfect insulation. He believed there was no pain, but added this reservation: "It is impossible to state positively whether the animal felt any pain. The immediate effect of the current is to paralyse the *motor* nerves . . . the effect on the sensory nerves cannot be stated." He concluded "Although there is no reasonable doubt in my mind that the killing as now carried out is painless, yet for this to be placed beyond all doubt . . .

would entail the design of fresh apparatus and considerable animal experimentation."

It is greatly to the credit of the R.S.P.C.A. of those days, and especially of Mr Paddison, that in spite of this generally favourable report they decided to give effect to the lingering doubt, and at Hopwood's suggestion they called in A. V. Hill; this in spite of anti-vivisectionist opposition and of the loss of face which would ensue if Hill's report should be unfavourable. Meanwhile I had been encouraged to keep up the pressure by, among others, that great-hearted old lady Miss Ada Cole, who was then fighting against the traffic in worn-out horses. She told me "I would like to back you openly, although I am an anti-vivisectionist, but my first duty is to my horses, and if I came out on your side my campaign would lose many of its anti-vivisectionist supporters", or words to that effect. The difficulties faced by Mr Paddison may be guessed from what passed at the Annual General Meeting in 1929. I had said (page 232 of the printed report) "There are four electrocuting machines used by the Society, one of which is used on very large numbers of animals. Now it is by no means certain that that method is humane. In fact, it may be cruel; it is difficult to say; and what I want to suggest is that, until we know, the method should be discontinued." The Chairman, Sir Robert Gower, then announced that "experts had been called in who were going to advise anew" but added, "I am going to say this, that we, as members of the Council, are not prepared to allow experiments to be made on animals" (applause).

A vivisection licence was nevertheless taken out for the premises. A. V. Hill, Lovatt Evans, Spilsbury, Rankine, and three veterinary surgeons all took part in the investigation on various occasions, but practically no fundamental scientific knowledge on the subject then existed. Hill did the only thing that was likely to yield information at short notice ; he switched on the current for one or two seconds and then examined the state of the animals. Some howled pitifully but most appeared to be dead and were presumed unconscious. (We should now infer that they were inert as a result of extreme muscular fatigue, but there were no grounds for such an inference at that time. The E.E.G.* had not been invented.) The results were variable and inconclusive. Mr Paddison gave his impressions to Captain Fairholme, then Chief Secretary of the R.S.P.C.A., in a letter dated 31st July, 1929. They were to the effect that Professor Hill agreed with him that " although no proof had been furnished of unconsciousness in the cats, there was no proof that they were conscious " (*sic* : he probably meant the converse) and " if the cats were conscious when exposed to the current the pain would probably be very severe ". There are gaps in my sources of information at this point, but it seems clear that in the end the Council of the R.S.P.C.A. very courageously and rightly decided to discontinue electrocution of cats and dogs at that time.

Whether any other users of electrocution followed this example I do not know, but it is certain that, in 1937, at least one electrocution cabinet made by the same manufacturers was in use by the Canine Defence League.

THE CAUSE OF DEATH

Meanwhile important progress had been made in American researches on the cause of death in electrical accidents. We in ULAWS followed them with keen interest and entered into correspondence with some of the scientists concerned, notably Kouwvenhoven and Ferris. Further, I had got into touch with Dr Foveau de Courmelles in Paris, and

* In the electro-encephalograph (E.E.G.) plates are placed on the head so as to pick up fluctuations of potential in the cortex of the brain. These are recorded. Their form is closely related to certain mental states.

also, by means of letters in the electrical press, had collected some more information about subjective experiences in electrical accidents. This was published in 1935 in Vol 3 of the *Animal Year-Book*, together with a report on Hertz's work at the Rothschild Hospital in Paris in which he used moribund human subjects as experimental material. We tried to keep the question open by publishing all such information, and writing letters in the *Animals' Friend* and elsewhere.

On 13th November, 1935, we had a letter from the Chairman of the Euthanasia Committee of the National Veterinary Medical Association (now British Veterinary Association) to the effect that the Committee was about to study electrocution. I have no record of what happened, but I feel sure we furnished all the information in our possession. The trouble was that it was all inconclusive at that time. It was sufficient to raise theoretical doubts but insufficient to prove that electrocution by the method then employed was cruel while appearing to experienced practical men to be entirely humane. However, in a letter published in the *Veterinary Record* of 7th November, 1936, a veterinarian, Mr S. J. Motton, described an electrical accident which had happened to himself. He said, *inter alia*, "If these animals feel anything of the sort experienced by me, then 'terrible' is about the only word which can describe it. I cannot believe it is easy to estimate the degree of suffering experienced by an animal tetanized by electricity, and very strong and convincing evidence should be produced before this method of preventing movement is accepted as painless."

Nevertheless, the Euthanasia Committee decided to sanction electrocution. In fairness to them it must be admitted that the wider a man's clinical experience, the more strongly convinced he must have felt by the appearance of unconsciousness in the animals, as judged by the criteria which are applied in chemical anaesthesia. This appearance has now been proved to be illusory, but it must have been hard to disbelieve interpretations of clinical signs when these interpretations had been vindicated in extensive experience with anaesthetics.

The most important result, for our purpose, of the American researches was to show that, under the conditions prevailing in the cabinets then used, the cause of death was heart failure due to ventricular fibrillation. When a heart is in ventricular fibrillation the ventricles operate irregularly and independently of each other so that the circulation of the blood is brought to a standstill. (It must be remembered that the current passed through the body only, missing the brain.) Two questions therefore arose: (1) How much time elapses, after the current has been switched on, before heart failure occurs? And (2) how long does consciousness endure after the incidence of heart failure? On the second question no definite information was available in those days, but happening one day to meet Professor E. B. Poulton I learned from him that he had once had a patient who suffered from intermittent ventricular fibrillation (reported in *The Lancet*, 30th November, 1918, p. 738). Poulton's impression was that in this case consciousness lasted only 2 or 3 seconds after each onset. We now know that the average time is considerably longer, actually about 12 seconds in the cortex and longer at a lower level, but for the time being I accepted Poulton's estimate provisionally and concentrated accordingly on question (1) above. Now if we knew the strength of the current, which could easily be measured, we could estimate, from the American results, how long current would have to flow before the heart failed. In March, 1937, I therefore asked Mr Charles R. Johns, then Secretary of the Canine Defence League, to allow us to measure the current in the GECAM cabinet used by them. He replied on 22nd March, 1937: "We have taken expert advice on this and are informed that there might be some danger to the man manipulating the installation, and if this be so I do not think the results would justify any risk . . . We judge . . . by the results we obtain, and physiological tests by experts have confirmed us in the choice of this method."

Meanwhile, on 14th April, 1937, Mr Paddison of the R.S.P.C.A. had written to me as follows: "Up to the present we are not exactly recommending electrocution, but we are considering the matter, as we have very good reports of the electrocution cabinet of the General Electrical Construction and Maintenance Ltd." Mr Paddison ended his letter with some very friendly and co-operative remarks, but what he had written was alarming. I therefore sent to him and to the Chief Secretary of the R.S.P.C A. (then Captain Fergus MacCunn) advance proofs of the article on "Electrocution of Cats and Dogs" which was about to appear in Vol. 4 of our *Animal Year-Book*. This article summarized the results of the American work, but it concentrated on the need for ensuring the use of current strong enough to cause prompt heart failure; I was still relying on Poulton's tentative estimate of the subsequent duration of consciousness, an estimate which, for the little it was worth, was the only information available on that aspect of the subject at that time. All these efforts were unavailing. In the *Veterinary Record* for 19th June, 1937, there was an exultant letter from the manufacturers; the Euthanasia Committee had found in favour of the GECAM cabinet.

HOW LONG DOES CONSCIOUSNESS SURVIVE HEART FAILURE?

In the autumn of that year I had a letter dated 28th September, 1937, from L. P. Ferris, of the Bell Telephone

Laboratories, New York, who had done what was, from our point of view, the most important of all the American researches on the causes of death in electrical accidents. He referred to my *Year-Book* article and took me to task for a statement which I have greatly regretted since. My only excuse for having made it is that we were feeling our way under difficulties, and had only Poulton's estimate to guide us provisionally as to the duration of consciousness.

Ferris said " I note that you say ' Our object must, therefore, be to ensure that electrocution apparatus shall be so designed as to produce fibrillation in the case of every animal.' I am sorry to have to question your conclusion, but I do not believe that the production of fibrillation is the most humane way of killing an animal by electric shock. . . There is a brief period . . . from the onset of the shock until unconsciousness sets in, during which an animal or a human being receiving a shock which causes fibrillation may suffer rather severely. I believe, from consultation with my medical associates in this work, that this period may last for a number of seconds. I take it that you would agree that the most humane way of killing an animal would be to avoid any cause for fear in the preparations and to render the animal unconscious in the briefest possible time. It would seem that a shock through the brain sufficient to instantly paralyse the central nervous system would be the most humane method."

Thus Ferris raised two important issues. (1) He held that consciousness is not destroyed within two or three seconds after heart failure, as we were provisionally assuming ; and (2) that if current passes through the head, it will cause instant unconsciousness. As to the second point a difficulty arose at that time, for simultaneously with our battle about electrocution we had on our hands another battle which must be recounted on some other occasion.

Here it must suffice to say that electric stunning, in which current is passed through the brain, had been introduced into the slaughterhouse without any understanding of the conditions requisite for producing electrical anaesthesia as distinct from what is now called " missed shock ", and we were fighting for a proper investigation of this topic. Hence we could hardly recommend electric stunning of dogs and cats until we had elucidated the conditions requisite for genuinely stunning pigs.

Ferris's other point, however, we took up at once. I wrote to my friend A. T. Phillipson, later Professor of Physiology at the Rowett Institute but then a post-graduate research student at Cambridge, and got back a remarkable, though admittedly tentative, estimate in a letter dated 6th February, 1938. This estimate started from the fact that consciousness will cease when the brain has used up its reserve of oxygen, and it was based on measurements of the rate at which blood gives up oxygen to the tissues of the body, in combination with measurements of the lowest partial pressure of oxygen in alveolar air which is compatible with life. Phillipson said " the fallacy of all this is, of course, that different tissues use oxygen at different rates and I can find no figures for the rate at which nerve cells use it. They are popularly supposed to be very sensitive to anoxaemia but I can't find any definite evidence on the point. However, I suggest the time of 20 seconds as the limit during which consciousness persists after the cessation of circulation, probably it is much less." Joyce Moore, afterwards Mrs Witt, who was on the staff of UFAW and had taken physiology for her degree at University College, London, discussed this letter with two other physiologists whose names I do not know. One thought that Phillipson's time was an over-estimate and that consciousness must fail immediately when the heart fails. The other physiologist came to the conclusion that " a matter of 10-15 seconds" was a reasonable guess. It is remarkable that these estimates, made so long ago by Phillipson and another physiologist by indirect inference from scattered data, agree so nearly with direct measurements which have recently been made by electro-encephalography.

A correspondence which took place at this time throws a rather interesting light on the evolution of our policy on electrocution. Just before hearing from Ferris I had received an enquiry from a Mr Charles Baldwin, a member of the Committee of the Manchester Dogs' Home, who were being pressed to adopt electrocution. In a letter dated 29th September, 1937, to which he sent a most cordial reply, I advocated taking steps to ensure a rapid arrest of the heart by ventricular fibrillation, but on 27th October, 1937, I wrote to him in these terms : " I have had a rather disconcerting letter from Professor Ferris of New York in which he suggests that stoppage of the circulation may not produce instant unconsciousness. I am pursuing this question further and will let you know the result . . . Meanwhile we can at least prevent the certainty of torture which arises when the current does not reach the prescribed strength." In view of the position in Manchester we decided to print and circulate a statement in the form of a letter dated 25th February, 1938, and addressed to Mr Baldwin, who was co-operating valiantly.

Here are some extracts from this printed letter. " In the first place, I think we ought to wipe out of our minds all such amateur tests as the appearance of the dog and its movement or absence of movement. An animal may exhibit violent reflex movements while it is totally unconscious. On the other hand, under electrical treatment it may be suffering intense agony while it is unable to move a single muscle. We should surely be acting wrongly if we decided to ignore the positive and unquestionable knowledge which has resulted from researches conducted in America for the purpose of ascertaining the cause of death in electrocution." After referring to asphyxiation, the letter went on thus: "Where death occurs from fibrillation, the duration of consciousness is a difficult problem. So long as consciousness does endure, the animal will be suffering intensely on account of the extreme contraction of the muscles . . . The question is, how long does consciousness endure after the circulation has been stopped by the onset of fibrillation? It depends on the reserve of oxygen available to the brain and can only be deduced by inference. One physiologist whom I have consulted puts it at 20 seconds, but another physiologist considers that this is an over-estimate and that the nerve cells must cease to function very rapidly if not immediately. A third physiologist puts the period at 10 to 15 seconds. Summing up, therefore, we must say that currents *below* the critical values which I have given in the *Year-Book* are known to cause death in a very painful manner. Currents above those critical values *may* kill humanely but *may* cause suffering which lasts for a certain number of seconds. As to this latter point, knowledge is not available at the present time."

In Vol. 5 of our *Animal Year-Book*, nominally dated 1938, there were two articles on electrocution. One, by an eminent veterinary surgeon, favoured the method upon which we were casting doubt, but did so on the basis of clinical observation and passed over the scientific evidence which I had collected. It may seem odd that an article which ran so contrary to UFAW's policy should have been published by UFAW, but we have always taken the view that responsible authorities must be allowed to say their say, even when we consider them mistaken. The second article, entitled "The Case for Suspending Judgement in Regard to Electrocution and Electric Stunning", was under my name. It said "In my previous article I assumed that fibrillation causes immediate loss of consciousness, but this conclusion has since been strongly challenged by one of the leading American investigators of electrocution . . . Physiological correspondents have tried to make an estimate based on the scanty available data, and have concluded that consciousness cannot be assumed to endure for more than about fifteen seconds at most after the onset of fibrillation, and that the duration may be much shorter than this. A sure answer does not appear to be available in the present state of knowledge."

At the end of 1938 Mr Paddison made a report to the Humane Slaughtering Committee of the R.S.P.C.A. in which he said that he had had a head electrode designed, and took the view that the results were very promising. This method, which seems to have been suggested to him independently by Professor Ashford of Bristol, was not, however, adopted.

EFFECTS OF THE SECOND WORLD WAR

With the outbreak of war the animal-protection societies were confronted with a difficult problem. As a result of first the evacuation and then the bombing of large towns, enormous numbers of unwanted cats and dogs had suddenly to be destroyed, and the most practical way of doing this unpleasant work was to use electrocution. Hence, when peace came and UFAW was able to resume its campaign for a scientific approach to the electrocution problem, conditions were extremely unfavourable for re-opening the question. It is a fact that the strength of a man's convictions is proportionate not to the evidence in favour of them but to the extent to which they have become habitual; this is extensively true. The societies using electrocution had been using it now for a number of years, on enormous numbers of animals, in the firm conviction that they were acting humanely, and meanwhile Mr. Paddison, one of the acutest and most devoted friends that animals ever had, had brought his years to an end as a tale that is told. It soon became clear that nobody from UFAW would be granted access to any of the electrocution cabinets which were in use by several societies in England. We tried our hardest to persuade the users to apply their minds to the facts which had been brought to light by the American researches, but we were entirely unsuccessful.

In fairness it must be recognized that the resolute opposition which appeared to us as obscurantism was in fact based on a sincere conviction that electrocution as then practised was humane and that UFAW was perversely trying to make a nuisance of itself out of mere cussedness.

On the other hand, certain societies have always generously accorded to UFAW the status of backroom boys which we have tried to merit. These are in particular the Scottish S.P.C.A. and the Ulster S.P.C.A.; nor must we forget our loyal friends the Equine Defence League. The first-mentioned two refused to adopt electrocution pending the production of scientific information as to how it could be made humane. The last-mentioned was the first society in this country to adopt the method mentioned above as having been recommended by Ferris. Later, too, the Manchester Dogs' Home was persuaded by Dr Dobbie to do the same. The P.D.S.A. has never adopted electrocution.

UFAW STARTS RESEARCH

However, UFAW emerged from the Second World War with enough accumulated funds to sponsor some research on its own. In 1948 we found in E. O. Longley, then on the staff of the Royal Veterinary College at Streatley, a scientist of brilliant intellect who was one of those rare birds, a man who has a clear grasp of the principles of both electrical science and veterinary science. Unfortunately he suffered from ill-health, but when he became available we decided to take a chance on that, and persuaded Professor Golla at the Burden Neurological Institute, Bristol, to provide facilities for the research we had in view. Later the enterprise was passed on by Professor Golla to Dr Derek Richter, who harboured us at the Neuropsychiatric Research Centre, Cardiff. The two institutions have a vast fund of knowledge gained in connection with the treatment of mental ailments in human beings by means of electricity, and the advantage of working with experts in that subject was for our purpose inestimable. Longley had to leave us on account of ill-health before he could obtain any practical results, but not before he had given us some valuable ideas and notably that of the c.p.r.* test for consciousness, which was later validated by Dr Croft.

Dr Phyllis Croft was introduced to us by Dr Richter when Longley left us. Her first task was to ascertain the conditions under which genuine anaesthesia can be caused by passing a current through the brain, and to devise a simple test for consciousness which, unlike the reflex tests usually used in chemical anaesthesia, would be valid in an electrified animal. Professor Golla gave us the clue. In all his experience he had never come across an instance in which an electroplectic fit, such as is used for treating certain mental ailments, had failed to produce instant unconsciousness, if the patient's own report could be relied upon. Dr Croft therefore studied electroplexy in non-human mammals, to see whether there also it invariably causes unconsciousness. One source of doubt was the fact that electroplexy gives rise to retrograde amnesia, so that patients' reports are not fully conclusive. Dr Croft

* Cardiac pain reflex.

used the electrocortigraph and the c.p.r. test to compare electroplexy with drug anaesthesia, and her results were published in the *Journal of Mental Science*. The most useful of them for our purposes was the validation of electroplexy as a test of unconsciousness.

We were now prepared to tackle the subject of electrocution as then practised: *i.e.*, with the GECAM cabinet, in which current passed from a collar on the neck either to a shackle electrode on a hind leg or to a metal plate on which the animal stood. Thus the current did not pass through the brain. (I refer to the "GECAM" cabinet for convenience, but this name, though usually employed, is not strictly correct. Originally, like many others, I had supposed that the apparatus was made by the General Electric Company in view of a resemblance between the trade marks. But one day C. C. Paterson, Director of Research at the G.E.C., put me right on this point, and he then took the matter up with his patent department. In a letter dated 31st May, 1938, the latter wrote to me as follows: "We have now taken up with the General Electrical Construction and Maintenance Ltd., the question of their use of the word GECAM as a trade mark in respect of their electrocution cabinets. Their reply is to the effect that they are now giving up the use of this word so far as their electrocution cabinets are concerned." The G.E.C. itself has never made such cabinets.)

In view of the known fact that death under the relevant conditions is caused by heart failure due to ventricular fibrillation, two questions had to be answered by the investigation on which Dr Croft was now embarking: (1) How long elapses after the current has been switched on and before the circulation of the blood ceases? And (2) how long elapses after the cessation of the circulation and before consciousness disappears? The first of these questions was attacked first, as being the easiest and not having to be studied on licensed premises, for all that had to be done was to measure the current flowing; given that, we could infer from the American measurements the time taken to induce fibrillation. Knowing that the contact resistance would fluctuate, we accordingly invested in an expensive duplex recorder, which was made by Evershed and Vignoles and adapted for us by Dr C. N. Smyth. The Royal Society lent us an Avometer.

AND RUNS INTO ROUGH WATER

It is easy to be wise after the event, and I can see now that we ought then to have taken the bull by the horns, purchased a GECAM cabinet, purchased dogs from one of the dealers who supply laboratories, and carried out the work on premises licensed by the Home Office for experiment. But in view of the fact that hundreds of dogs were being electrocuted every week in these cabinets we looked for an easier way out, and always one seemed to be on the horizon. As a result we lost a great deal of time. Several societies toyed with the idea of allowing us to make measurements in their clinics, but eventually refused absolutely to do so.

The attitude of the R.S.P.C.A. was clarified at a talk between its Chief Secretary and myself. That society was unwilling to accord to UFAW the status of backroom boys to which we aspired, and entrenched itself behind the authority of the British Veterinary Association. This gave me an idea. A lot of water had flowed under the bridge since the B.V.A. had gone into this matter before, and they might be willing to re-open it in the light of the new knowledge now available. And so it turned out, and it is thanks to the open-minded attitude of that Association that the facts have at last been recognized.

I was allowed to write in the *Veterinary Record* (23rd December, 1950) an article in which the technical data then available were summarized, and eventually we arranged with the B.V.A. that Dr Croft should work at an R.S.P.C.A. clinic under their aegis, though all expenses, including her salary, would be paid by UFAW. In this way the refusal to co-operate with UFAW was circumvented.

I don't think anybody was really to blame for the exasperating delays which then occurred, unless, perhaps, it be the firm which did the wiring for us and was rather slow off the mark. The last place in the world for carrying on scientific research is a busy clinic where one's activities are a real nuisance (although, on the other hand, Dr Croft's veterinary qualifications enabled her to render some uncovenanted help which seemed to be appreciated). In view of the fact that a pressure of 2,000 volts was being used, the leads to the recorder had to be brought out in conduits to satisfy Board of Trade regulations. We were trying to measure the current through the body, but in this particular cabinet part of the current passed down the forelegs and so missed the trunk, and as a result the construction had to be modified so as to enable the requisite measurements to be taken. Then unforeseen reactions in the animals necessitated further modifications. To lay folk who had not the faintest idea of what the scientists were driving at we must have seemed about as useful as a handful of spanners thrown into the works of a printing press.

THE SURVIVAL OF CONSCIOUSNESS ONCE MORE

Things were at this stage when my friend Edgar Dand had the misfortune to develop heart block, and mentioned to me that on one occasion his heart had stopped for 15 seconds, as timed by his watch. The relevance of this observation struck us at once. It showed, so far as it went, that consciousness can endure for at least fifteen seconds when the supply of blood to the brain has ceased. Dr Croft at once embarked upon a library research, with sensational results. She unearthed three papers which fully supported the above conclusion, including a particularly important one by Sugar and Gerard. (It is a curious fact that these researches seem to have begun at just about the time when we were first asking physiologists our questions on the subject, and I wonder if this was the cause of that.) Although the times given in these papers had a considerable scatter they were much the same as those which Phillipson had predicted before the war from purely theoretical considerations. Their time for the cortex was twelve seconds, which agrees with Roberts's later and corresponding measurements. Their times for the thalamus, with which an animal can probably appreciate but not locate pain, are longer than this.

It goes against the grain to stop a research in mid stream but there seemed to be no point in going on with the measurements we were then engaged upon. Even if the GECAM cabinet caused instant fibrillation, these animals would suffer for a long time, of the order of twelve seconds or more, after the heart had failed. It seemed ethically unjustifiable to go on electrocuting, and thereby causing pain, merely in order to find out whether the time might be still longer than that, and Dr Croft reported accordingly to the Euthanasia Sub-committee of the B.V.A.

We have always been anxious to avoid sensational publicity, for fear, in particular, of prejudicing the work of the R.S.P.C.A. inspectors, which is the backbone of the protection of animals in England. We intended that Dr Croft should announce her finding in a public lecture on 28th November, 1952, but that users of the cabinet should be notified in good time, so that by the date of the announcement they might be in a position to say that they had modified their cabinets by the simple device, which we recommended as an interim measure, of substituting an ear-clip electrode for the collar electrode; this would have afforded a great improvement though it would not have made the process infallible. Unfortunately Chapman Pincher got wind of what was going on and published a sensational headline in the *Daily Express*. We have never discovered who let the cat out of the bag, but anyhow our advice to modify the cabinets provisionally was not taken even then by the societies which were using them.

THE B.V.A. INSTITUTES A PARALLEL RESEARCH

A meeting of the Euthanasia Sub-committee of the B.V.A. was called for 3rd December, 1952, and that evening I wrote to Lord Merthyr, Chairman of the R.S.P.C.A., as follows: "Mr Moss will be informing you that the Euthanasia Sub-committee of the B.V.A. decided this morning to have the existing cabinet investigated by specialists in electro-neurology nominated by itself, and accepted the R.S.P.C.A.'s generous offer of financial help. This is substantially what I have been fighting for, *Athanasius contra mundum*, for the past 30 years, and by the irony of fate it comes just when it seems, to Dr Croft and myself and our associates, to have been rendered superfluous by new information which has just come to hand. However, it would probably be unreasonable to expect our *ipsi dixerunt* to be accepted just like that, and the middle course which has been decided upon is probably the best that we could reasonably hope for."

Nevertheless we dissociated UFAW from this parallel research on the ground that further experimental electrocutions would entail unnecessary suffering, since in our view our case had been proved. The R.S.P.C.A., on the other hand, must have still believed that the process would be proved to be humane, for besides financing the research they continued to use the GECAM cabinets in the interim. The physiologist appointed to do the work was Dr T. D. M. Roberts of Glasgow, and, like most serious researches, it took a good deal longer than had been foreseen.

THE DEVELOPMENT OF THE ELECTROTHANATOR

While awaiting confirmation of our findings we turned our minds to the development of a satisfactory substitute for the existing cabinet. Our starting point was the technique adopted by Dr. Voûte, a Dutch veterinary surgeon, who simply put a crocodile clip on the right ear and another on the left flank and connected the clips to the mains through a foot-operated switch. There was still the problem of obtaining dogs for tests, but Mr. Hodgman, of the Animal Health Trust, having occasion to destroy eleven dogs, kindly afforded facilities to Dr Croft; and using electroplexy as the sign of unconsciousness she made the investigation reported in *The Veterinary Record* for 25th April, 1953. So far as it went it justified the method, but the sample was small and narrow, the dogs being all of two breeds.

It was at this stage that Dr Vinter became active in the development of a fully satisfactory apparatus. While visiting North Africa on behalf of the Society for the Protection of Animals in North Africa (SPANA) she became concerned with the problem of stray dogs, which was urgent there.

In November, 1953, she tried out the Dutch technique in Rabat, but found that with the mains voltage there available it took 2, 3 or 5 seconds to produce a fit, and she considered that this was too long in view of the severity of the pain entailed.

On return to England she contrived to get into touch with Mr Angus MacPhail, one of the foremost designers of electromedical equipment, who at once applied his outstanding ingenuity and knowledge to the task. The apparatus now known under the registered trade mark "Electrothanator" was thus evolved. Its main features are (1) the electrodes, specially designed for elimination of various defects which are exhibited by crocodile clips; (2) the circuit, which ensures a current through the head strong enough to stun instantly, together with a current through the heart which kills almost as rapidly; (3) a continuity-tester, which indicates beforehand whether the circuit is in fit condition for switching on; (4) an automatic time switch, which ensures that electroplexy shall be verifiable if present; and (5) precautions against the harmful effects of slovenly maintenance, which have been all too evident in the past.

Thus it came about that when the B.V.A. were in a position to disclose the conclusions reached by Dr Roberts, we were ready with an alternative apparatus which had been perfected, or almost perfected, by close collaboration between Dr Vinter and Mr MacPhail. The meeting was called for 9th June, 1954, and Dr Croft and I were allowed to take with us Dr Vinter, Mr MacPhail, and an Electrothanator which had been made to the order of SPANA.

THE CRUELTY OF THE OLD METHOD CONFIRMED

Dr Roberts's investigation had confirmed, by direct electro-encephalographic observations, the conclusion which we had previously reached by indirect inference; even to the time, averaging twelve seconds, during which cortical indications of consciousness survive the onset of fibrillation. In one respect he went even further than we had gone. After stimulation for a few seconds an animal is found to be flaccid and inert when taken out of the cabinet. The

American researches did not explain this puzzling phenomenon, and I had tentatively (but erroneously) suggested that the animal might have swooned from pain shock (not a good way of causing anaesthesia). Dr Roberts found, however, that consciousness still persists even during this inert phase, and he concluded that the muscles must have been put out of action by extreme fatigue although the brain was still functioning.

In one instance Dr Roberts switched on the current for only half a second, a time too short to put the dog out of action. The severity of the pain caused by this method of electrocution was proved by the fact that the dog went on howling for 26 seconds, and thereafter showed intense fear of the operator. Dr Roberts made a gramophone record of the howling and thus provided evidence which has convinced some sceptics.

THE PRESENT POSITION

Mr Gelder, the new Secretary of Our Dumb Friends' League, immediately offered full co-operation and soon had an Electrothanator installed. It has been working satisfactorily ever since. The Battersea Dogs' Home were not far behind; and the Equine Defence League, who had long been using a head electrode in their Carlisle Dogs' Home, decided to adopt this further improvement. In November, 1954, at a further meeting, the Euthanasia Committee of the B.V.A. laid down standards to which all electrocution cabinets must conform if they are to be regarded as humane. The British Standards Institution then appointed a committee, on which UFAW was represented, to draw up a specification and this is now available. The R.S.P.C.A. decided that all electrocution cabinets used in its clinics must conform to it. It is conceivable that some GECAM cabinets in other hands may still be in use, but we do not know of any and it seems probable that victory is complete so far as Great Britain is concerned.

Unfortunately, however, this is not true of other countries. One of the leading electrical firms in Germany has put a very cruel electrocution cabinet on the market and refuses to listen to reason. It is extremely hard to make foreigners understand the technical facts, and outside this country the battle is scarcely half won.

Further, there is no humane method of electrocuting cats and probably there never will be one.

The Electrothanator is a scientific instrument, and scientific instruments cost money. Provided the price be fixed on an accurate basis of costing, we feel that nothing short of the best is acceptable for animals, in view of all the suffering that has been inadvertently inflicted on them in the past. No risks ought to be taken by trying to save money with inferior materials, workmanship, or design. All of these should be of the highest quality used in the highly specialized field of electromedical apparatus.

In no circumstances will UFAW or any of its officers accept any financial interest in apparatus for electrocution. Our sole concern has been to see a reliable machine made available.

In Britain this long struggle has ended very happily; everybody who matters now accepts the facts and acts upon them, and the work which brought them to light has been recognized by two very gracious acts. The Battersea Dogs' Home voted a gift of £100 to UFAW in recognition of its work, and the Royal College of Veterinary Surgeons awarded to Dr Croft and Dr Roberts the Livesey Medal, which is given from time to time to the person who "has done the most serviceable work towards the prevention and/or alleviation of pain and/or fear in animals".

In Praise of Anthropomorphism

Read before the Association for the Study of Animal Behaviour at Birkbeck College, 9th January, 1959, and Reprinted from *The UFAW Courier* No. 16, Autumn, 1959.

THE WORD " anthropomorphism " means, in this paper, the practice of using one's own subjective experiences as a method of representing to oneself, by analogy, the subjective experiences of other species of animals. The word " automorphism " may be similarly used to denote the practice of using one's own mind, by analogy, to represent the minds of other human beings.

Some people, such as Konrad Lorenz, Hediger, and Len Howard, have an intuitive understanding of animals. Others lack that gift, and may even be led to suppose that nobody else can possess it. Moreover every analogy breaks down if it is pushed too far or is carelessly applied, and the analogy called " anthropomorphism " is no exception. For these two reasons anthropomorphism is under a cloud at the present time, and my purpose is to point out that the reaction against it has gone too far ; that, in fact, what began as scientific criticism has degenerated into quasi-philosophical pedantry.

A reasonable view of the animal mind has been expressed by Sir Julian Huxley and the Wellses in the following terms (1937) : " It is a curious fact that the scientific mind and the activities of the reasoning faculty are so frequently written down as ' inhuman '. Actually this cold power of abstraction, this ' inhuman ' reason, is the one emergent property which the human species alone possesses, while our warm human emotions we share with the brutes. There can be no reasonable doubt that other mammals are subject to the same kinds of passions, feel the same sort of emotion, as we ourselves. Our sheep can be frightened; our dog is glad when we come home, feels something closely akin to shame when caught in some misdeed; our cats can experience anger and disappointment. But the capacity to subtract eleven from twenty-four .... to attach any meaning to abstract terms such as space and truth—this is all distinctively and exclusively human." The words here quoted as denoting emotions—that is, the words "frightened", "glad", "shame", "anger", "disappointment"—are anthropomorphic; they are applied to animals on the assumption that states of mind roughly similar to what we observe in ourselves are to be found in other creatures also.

THE POSITIVE EVIDENCE

(1) In the first instance we impute such feelings to animals intuitively in the way in which we interpret the expressions on a human face, which are extremely difficult to analyse objectively (Darwin, 1889).

(2) This intuitive view of the animal mind is confirmed by the experience of persons who have to deal with animals in a practical way; of trappers and hunters who have to outwit their victims and of trainers who have to teach horses, domestic pets and circus animals. The fact that these people use anthropomorphic assumptions, and do so with success, is *prima facie* evidence that the assumptions themselves are sound at bottom.

(3) Anthropomorphism is *prima facie* justified in advance by the fact that the human race is akin to the lower animals through the process of evolution. Opposition to Darwin's theory, arising from contemptuous feelings towards the

lower animals, is dead so far as the descent of man's body is concerned, but a similar sentiment in respect of mental kinship seems still to exert an unconscious influence on many biologists. Admittedly the fact that animals do not use a grammatical language suggests that they are incapable of abstract reasoning, but reason is one thing and feeling is another. The capacity for sensation and emotion appears to have been a primitive, not a late development. By remembering the subjective experiences which we had when we were two or three years old, when we were less intelligent than now, we can see that our feelings then were more acute than they are to-day, and it may be that in this respect ontogeny recapitulates phylogeny.

(4) The behaviour of animals in response to traumatic stimuli is closely similar to that of human beings. (Grindley, 1933.) It consists of voluntary movements such as writhing and, in the case of species in which parental care obtains, screaming ; and of autonomic changes of pulse-rate, breathing and the like, as in man.

(5) The nervous structure of the lower animals is analogous to that of man (Baker, 1948). Nerve fibres functioning in the same way as sensory nerve fibres in man are incorporated in structures which are homologous in all the mammals, including man, and analogous, though decreasingly so, in species of decreasingly complex organization.

(6) Appetitive behaviour differs from mechanical reflexes by showing some degree of flexibility. It is only in part that the pattern of instinctive action is fixed. Birds adapt their nest-building to the details of nesting sites and materials that happen to be available. Even an ant, finding her way home, will adapt her course to circumvent obstacles.

(7) Many species have a language of some sort ; gestural in honeybees, vocal as well as gestural in most vertebrates.

(8) The electroencephalograms of mice, rats, rabbits, cats, dogs, monkeys, goats, sheep and cows have been observed. They are broadly analogous to those of human beings.

(9) Anxiety can cause psychosomatic effects such as gastric ulcer in dogs, rats and other animals.

(10) Animals can learn or be conditioned. Learning, as distinct from reasoning, is sufficiently similar in man and beast for the study of learning processes in rats, for instance, to have thrown light on learning processes in children, and to have been of value to the teaching profession.

Experiments on learning often take the form of conditioning by positive and negative reinforcement ; that is, if plain English is permissible, by rewards and punishments. In order to be conditioned in this way, the subject must be capable of two different things. He must be able to make a distinction between reward and punishment ; that is, he must be sensitive to pleasantness and unpleasantness. Secondly he must be capable of associating the signal with the pleasant or unpleasant experience which constitutes the reward or punishment. This ability to form mental associations we may rank as intelligence, and it is to be noted that intelligence alone will not make conditioning possible ; there must also be sensibility. Thus conditioning arises from a combination of two quite different factors, intelligence and sensibility. There is no evidence, so far as I am aware, that intelligent creatures are more sensitive than stupid ones, for the two factors are confounded in experiments.

Thus anthropomorphism is justified by our intuitive interpretation of animal behaviour, by everyday experience in dealing with animals socially, by the theory of evolution, by the way in which animals react to traumatic stimuli, by analogies in the structure of the nervous system, by flexibility in appetitive behaviour, by the existence of rudimentary language, by electroencephalography, by the occurrence of psychosomatic maladies, and above all by the ability of animals to learn. In the teeth of so much cumulative evidence a heavy burden of proof rests on anybody who would repudiate anthropomorphism.

ANTI-ANTHROPOMORPHISM

But there is a story of a countryman who, on staying at a London hotel for the first time, enquired about getting his boots cleaned and was told to leave them outside his bedroom door. He was not, however, to be so easily deceived. Thus forewarned, he slept with them under his pillow.

A similar fear of being deceived besets many ethologists, who, when they are obliged to refer to hunger or pain in an animal, write these words between quotes to be on the safe side. Professor Pumphrey (1952) has pointed out that "when the accurate and proper use of language has entrapped a zoologist into a statement that seems to him heretical, it is quite usual to hear him apologize for speaking teleologically, and he generally looks as sheepish and embarrassed about it as if his bedroom had been found full of empty whisky bottles". Professor Pumphrey adds that a language has accordingly been devised which is designed to conceal such inconsistencies from those who use it. Thus "Man can see, but animals are photo-receptive. Man remembers, but we must not say that an elephant never forgets; to be objective we must say that elephants seem to behave as if they were subject to a persistent mnemic hysteresis resembling the behaviour of steel in an inconstant magnetic field. Man is capable of loyalty but a dog is conditioned by the association of his master's approbation with food. Man reasons from experience, animals learn by trial and error. Man is swayed by emotions, animals by 'action specific energy'".

Those who repudiate anthropomorphism in this way—let us call them "anti-anthropomorphists" for short—seem to do so on two grounds.

DIFFERENCES BETWEEN MAN AND BEAST

The first is the same as that on which anti-vivisectionists claim to discredit the use of animals for medical research, namely that there are wide differences between human beings and other species, and hence, it is urged, analogies between the two are invalid and misleading. Anti-vivisectionists point out that the physiology of a dog differs in many ways from that of a man, and that dogs and men react differently to various drugs and pathogens; and the argument which is used by anti-vivisectionists in relation to the bodies of men and beasts, is used by anti-anthropomorphists in relation to the minds of these species. Anti-vivisectionists are so afraid of being misled by a false analogy between man and beast that they abjure medical science; and anti-anthropomorphists are so afraid of being similarly misled, that they will not write of "hunger" or "pain" in animals unless they put these words between quotes (in defiance, be it noted, of one of the canons of literary style).

A fair answer to anti-vivisectionists and anti-anthropomorphists would seem to be this, that it is wrong to condemn the use of comparative methods merely because the analogies on which they depend can be pushed too far; and that analogy, though misleading if used uncritically, has, when used with due precaution, justified itself by results as an indispensable tool of scientific research.

Thus, to an anti-vivisectionist it may be pointed out that the cytology and genetics even of plants has thrown light on the cytology and genetics of human beings, and to the anti-anthropomorphist it may be pointed out that if a robin's aggressive behaviour is released quite irrationally (Lack, 1953) by certain releasers which include a red breast, natural or factitious, and a robin-like shape, we can get some insight into the subjective aspect of the bird's responses from analogy with the way in which in the human male the I.R.M. of courtship is released by scarlet lips, natural or factitious, and a Bardot-like shape; and in which in both robins and humans these primary responses can be modified or suppressed as a result of learning.

ANALOGICAL THINKING

Perhaps both anti-vivisectionists and anti-anthropomorphists confuse concepts with mental images, and hence misunderstand the nature of analogy. The distinction may be illustrated by the following example. A person with normal vision can form a mental image of the world as seen by a person who is totally colour-blind; this he can do by abstracting from differences of hue, and imagining a monochromatic picture of the world. But a totally colour-blind person, who sees monochromatically, cannot form a mental image of a polychromatic world; he can only conceive of it by analogy with the qualitative differences which he perceives between sensations of some other kind, as in the case of John Locke's blind man who supposed that the colour scarlet is like the sound of a trumpet, a very apt comparison. Again, I find it impossible to form a mental image of the feelings of a mother for a new-born baby, which to me is a repulsive and even a disgusting object. But if I say that women love these creatures, I am not obliged to write the word "love" between quotes, and I am quite able to attach a meaning to it. This word denotes a class of sentiments, some of which I can feel though not others. Concepts of this sort are formed, as Spearman

(1927) pointed out, not as a relation between two terms but as a second term when a first term and a relation are given. Such a concept as maternal love is a member of a class defined by the relation, and I can think conceptually of members of the class even when I cannot imagine them. Again, the amorous sentiments of men are almost certainly different from those of women, but this does not make amorous conversations meaningless, and in communications passing between the parties, even if these be ethologists, it is not, if my information is correct, considered necessary to write the key words between quotes. We can, indeed, use a mental image as a specimen of the class determined by a concept, " to fix our ideas " as we say. In a proposition of Euclid we can visualize a particular triangle as standing for all triangles, that is for the concept of a triangle ; and we can think of a particular experience of pain, such as a kick on the shins or a burn, as a symbol for the concept of pain. But it is in the concept, not the mental image, that the generalization resides, even when the image is a composite one.

DESCARTES, RAY, LOCKE AND DAVID HUME

Aristotle and Aquinas allowed to animals what they called a " sensitive soul ", as distinct from what they called a " rational soul ", but a notion that animals are automata without any minds at all was put forward in the seventeenth century by Descartes against the following background. Like many others, he confused Plato's doctrine of the intrinsic immortality of the human soul with the Christian doctrine on a kindred subject, and wishing to uphold it he wrote (1637) : " I have dealt at length with the subject of the soul because it is one of the most important ; for, after the error of those who deny God . . . there is none which more readily leads weak intellects away from the straight road of virtue than to imagine that the soul of an animal is of the same nature as our own . . . whereas when one knows how different these are, one understands better the reasons which prove that our soul is of a nature entirely independent of the body, and that consequently it is not liable to die with the latter." Thus Descartes had a reason for wishing to make as much as possible of the difference between men and animals. He noted that even in human beings the "animal spirits" have power to move the limbs, and so we see, he says, "how their heads, shortly after being cut off, still move and bite the ground in spite of being no longer alive". He adds that this "will seem by no means strange to those who, knowing how many various automata, or moving machines, the industry of men can make, using only very few parts, in comparison with the large number of bones, muscles, nerves, arteries, veins and all the other parts which are in the body of every animal, will look upon the body as a machine which, having been made by the hands of God, is incomparably better organized . . . And if there were such machines having the organs and external form of a monkey or some other animal devoid of reason, we should not have any means of knowing that they were not entirely of the same nature as these animals". Descartes then claims that, on the other hand, if automata existed in the form of men the difference would be detectable, first from the fact that men can converse and secondly because, instead of their reactions to stimuli having a rigidly fixed pattern, as he supposed the reactions of animals to have, "reason is a universal instrument which can serve for all sorts of situations".

These views were rebutted by the Cambridge platonists, Cudworth and Henry More, and by the naturalist, John Ray (1692; Raven, 1942, 1951), and that most level-headed of philosophers, John Locke, also rejected them. "Birds learning of tunes," he wrote (1689*a*), " and the endeavours one may observe in them to hit the notes right, put it past doubt with me that they have perception and retain ideas in their memories and use them for patterns. For it seems to me impossible that they should endeavour to conform their voices to notes (as it is plain they do) of which they had no ideas. For though I should grant sound may mechanically cause a certain motion of the animal spirits in the brains of those birds while the tune is actually playing; and that motion may be continued on to the muscles of the wings, and so the bird be mechanically driven away by certain noises, because this may tend to the bird's preservation; . . . it cannot by any appearance of reason be supposed (much less proved) that birds, without sense and memory, can approach their notes nearer and nearer by degrees to a tune played yesterday; which if they have no idea of in their memory, is no-where, nor can be a pattern for them to imitate, or which any repeated essays can bring them nearer to." However, Locke's commentator Dr. Morell (1793) objected that " in this instance of birds there is something wanting of convincing evidence that they are conscious of what they do. That the singing of some birds is in a great measure mechanical is manifest from their singing more briskly in a room where there is most walking, talking, or any sort of noisy motion ".

Descartes himself was not so dogmatic on this subject as his followers were, and he even wrote thus to Monsieur Morus (1649): " I am referring to thinking, not to life or feeling; for I do not deny life to any animal . . . I do not even refuse them feeling, so far as this depends on the organs of the body ". But what Descartes put forward as a tentative speculation was adopted as a ruthless dogma by his followers, and notably by the Port-Royalists (Sainte-Beuve, 1878). They vivisected animals for fun, without anaesthetics of course, to show their friends the circulation of the blood; and Malebranche used to amuse himself by kicking his dog in order to hear what he took to be the "creaking of the machine".

In the eighteenth century my distinguished clansman David Hume, on the other hand, did not accept these views. In the course of his acute analysis of inductive inference, which has set logicians by the ears ever since, he devoted a whole section of his *Treatise of Human Nature* (1738) to the subject of "The Reason of Animals", using the word "reason" to denote what we should call "learning" in addition to logical inference. Speaking of learning in animals, he says of the dog: "The inference he draws from the present impression" (*i.e.*, sensation) "is built on experience, and on his observation of the conjunction of objects in past instances. As you vary the experience, so he varies his reasoning. Make a beating follow upon one sign or motion for some time, and afterwards upon another; and he will successively draw different conclusions, according to his most recent experience". And earlier he says: "The resemblance between the actions of animals and those of men is so entire, in this respect, that the very first action of the first animal we shall please to pitch on will afford us an incontestable argument for the present doctrine"; the doctrine, that is, that "the beasts are endowed with thought and reason as well as men".

Nevertheless the Cartesian theory of the robot animal continued to encourage a contemptuous attitude which is still widespread on the Continent and is summed up in the formula "Animals have no souls".

CHARLES DARWIN

It was this habit of mind that underlay the opposition encountered by Charles Darwin, and with the eventual acceptance of Darwin's teaching the pendulum swung back in favour of anthropomorphism. Darwin's friend G. J Romanes (1882, 1883), in particular, sought to bring the development of mind within the ambit of his theory. The work of Romanes on animal intelligence resembled the work of Darwin on natural selection in being based, like astronomy, on careful observation and hard thinking with comparatively little controlled experimentation. Nevertheless when Romanes is scorned as a mere anecdotalist one may wonder whether the baby is not being emptied out with the bath water. He claimed that reflexes comprise only " particular adjustive movements in response to particular stimulations " and wrote of the " variable and incalculable character of mental adjustment as distinguished from the constant and foreseeable character of reflex adjustments ". (It is interesting that, writing three-quarters-of-a-century ago, he took account of the performance of calculating machines.) He pointed out that while the presence of flexible instinctive responses and of learning is evidence of mind in animals, the rigidity of reflexes does not prove the absence of mind but merely fails to prove its presence.

The pendulum swung back against anthropomorphism under the influence of Lloyd Morgan (1894), who introduced his famous canon that began as a rule of procedure and ended as a dogma. It read thus : " In no case may we interpret an action as the outcome of a higher psychical faculty, if it can be interpreted as the outcome of the exercise of one which stands lower in the psychological scale."

Lloyd Morgan thus deliberately imported a systematic bias into the interpretation of animal behaviour. It was as if you were to rule that, in a parliamentary election, all electors who did not go to the poll must be counted as having voted for the socialist candidates ; or that, at the Heligoland bird trap, all birds found without rings on their legs are to be reckoned as having come from Russia. Surely the right course to adopt in interpreting animal behaviour is to keep an open mind until the evidence is seen to favour one rather than another of the possible explanations that can be suggested for any particular instance of it; and meanwhile to adopt, provisionally, that explanation which best commends itself to common sense at the time.

Now the faculty which stands lowest in the psychological scale is that of showing invariable reflexes, and Lloyd Morgan's canon implies that if any behaviour can, with a little ingenuity, be explained away as a mere reflex, that explanation is to be favoured in preference to any other, in conformity with the arbitrary bias prescribed by the canon. It is accordingly to the widespread adoption of Lloyd Morgan's canon that we must, I believe, attribute the practice of writing such words as "hunger", "fear" and "pain" between quotes when they relate to animals, though never when they relate to human beings.

OBJECTIVISM OR POSITIVISM

So far I have been discussing the fact that the great differences which exist between man and beast have been used not only by anti-vivisectionists as an argument for rejecting comparative biology, but also by anti-anthropomorphists as an argument for rejecting anthropomorphism. I now come to a second manner of thinking which, if followed out consistently, must apply to man and beast alike and, if converted into a dogma, must issue in solipsism. I refer to what is called "objectivism" by ethologists and "positivism" by philosophers; that is, to the doctrine that nothing can be known except phenomena and the relations between them. I do not propose to criticize objectivism here, but merely to point out that it originated with human psychology, and can be applied to man and beast alike.

In the same way that anti-anthropomorphists refuse to interpret observations upon animals in terms of mind, Pyrrho of Elis refused to interpret any of his observations in terms of either minds or objects. " And his life ", says Sextus Empiricus " was consistent with this. Turning aside for nothing and avoiding nothing he took everything as it came, if chariots happened to meet him, and precipices, dogs and the like, allowing no weight to the evidence of his senses. But he was looked after (as Carystius Antigonus tells us) by his friends who followed him . . . And once when Anaxarchus had fallen into a ditch he passed by and lent him no help. And when many censured this action, Anaxarchus himself commended him for having borne himself unconcernedly and without emotion."

The fourteenth-century schoolman William of Ockham, whose name is familiar in connection with Ockham's Razor, pointed out that it is impossible to prove that any particular man *is* a man, and not a corpse manipulated by an angel after the manner of a puppet (Crombie, 1952). If you are a dogmatic objectivist and wish to be strictly logical and consistent, you must say " in quotes " every time you mention anybody by name, for it is impossible to prove that he is not an illusory phantom.

The word " positivism " was invented more than a century ago by the French writer Auguste Comte, who also elaborated the principle denoted by it and by its modern synonym " objectivism ". Although most of what Comte wrote was superficial, several pregnant ideas are to be found buried in the floods of trivial verbiage which fill the six volumes of his *Cours de Philosophie Positive*. His positivist principle is summarized thus by J. S. Mill (1865) : that " we have no knowledge of anything but phenomena ; and our knowledge of phenomena is relative, not absolute ".

Comte's other pregnant ideas to which I have referred include an insistance on the importance for psychology of brain physiology, which he (1838*a*) took to mean phrenology, and a forecast of the value of the comparative study of animal behaviour (1838*b*). But the point I wish to emphasize is this. Garrulous and inconsistent though Comte's writing was, and full of base-degrees (Jeffreys, 1957) fallacies, he did at least recognize that his principle must be applied to man as well as to beast if at all. Thus he wrote (1838*c*): " The naturalists have forced the metaphysicians to renounce at last the singular device thought up by Descartes, and to recognize, more or less explicitly, that the animals, at all events in the higher part of the zoological scale, manifest in reality most of our emotional faculties and even our intellectual ones, simply with differences of degree." And he remarks (1838*d*): " Doubtless a cat . . . though unable to use the pronoun ' I ', does not usually mistake itself for some one other than itself." He also denounced (1838*b*) " the vain and unseemly declamations of theologians and metaphysicians against the pretended tendency of such a doctrine to degrade human nature ".

Positivism was adopted, under another name, by J. B. Watson (1925), the founder of behaviourism. Watson considered that the work of all the psychologists who had preceded him, such as Wundt, William James and MacDougall, was worthless, and he wrote contemptuously of " the major assumption that there is such a thing as consciousness and that we can analyse it by introspection ". In this state of unconsciousness he proceeded unconsciously to write a course of lectures for the edification of his unconscious hearers and readers, but at least he applied to the human mind the same negations that he applied to the animal mind.

W. H. THORPE

It is pleasant to turn to a more judicious and balanced writer. W. H. Thorpe (1956) has this to say in the course of a reference to Sommerhoff: "Sommerhoff's arguments have shown, perhaps more dramatically than ever before, that *both* subjective and objective concepts are necessary in the scientific study of life, and that a biology which succeeds in being purely objective, however powerful it may be in parts of its field, must fail to provide a full biological philosophy—just as surely as a purely subjective biology would be unthinkable. In other words, subjective concepts derived from introspection are among the essential tools for the study of life, and the only hope for a complete biology is to combine subjective and objective in the right proportions, using each approach with due circumspection and adequate safeguards as it is required."

KONRAD LORENZ AND N. TINBERGEN

With these wise words in mind, let us turn to the Lorenz-Tinbergen school and say in their praise that they use objectivism, that is to say positivism, as a method without converting it into a dogma. On one hand Lorenz devised such mechanistic concepts as reaction-specific energy, innate releasing mechanisms, and imprinting, and Tinbergen (1951*a*) specifically objects to the incorporation of any consideration of subjective phenomena or directiveness into a study of the causation of behaviour. These workers are seeing, in fact, how far they can get with the mechanistic hypothesis of a hierarchy of instincts governed by a hierarchy of neural mechanisms, and find the method a fertile source of ideas for experiments. On the other hand *King Solomon's Ring* (Lorenz, 1950) is frankly anthropomorphic, and illuminatingly so, and as Tinbergen nears the end of his *Study of Instinct* (1951), and still more in his *Social Behaviour* *in Animals* (1953), he reminds me a little of Dr. Johnson's friend who had tried in his time to be a philosopher but found to his dismay that cheerfulness would keep breaking in; in Tinbergen's case, subjective insights will keep breaking in. In chapter VII of his *Study of Instinct*, Tinbergen explicitly faces the task of integrating the subjective and objective interpretations of animal behaviour, though he might prefer a different way of saying this. I must confess that his synthesis takes me a little out of my depth, but Professor C. A. Coulson (1955) compares the parallel problem, of synthesizing the scientific and theistic views of the universe, to that of mentally synthesizing a plan and a sectional elevation of a building. It is easy to understand either drawing alone, but a greater intellectual effort is needed to synthesize the two.

I suspect that an ethologist would not know what components of behaviour to look for if he did not, consciously or unconsciously, use subjective clues. I have already referred to the analogy between our intuitive interpretation of animal behaviour in anthropomorphic terms, on one hand, and our intuitive interpretation of the expression on a human face, on the other. Now Darwin (1889*b*) reproduced some photographs, taken by Duchenne, of expressions induced by electric stimulation of an old man's facial muscles; but could Duchenne have found the right muscles to stimulate if he had not followed a subjective clue? And could Leonardo da Vinci have painted La Gioconda with the right expression, but for the clue afforded by his subjective interpretation of her smile? I suspect that, in the same way, an ethologist must look over his shoulder at his subjective insights into the meaning of behaviour, for otherwise he would have no clues to suggest the mutual relevance of the various components of a behavioural syndrome, and to help him, provisionally at all events, to group those components under the right instincts.

And then, having looked over his shoulder at his illicit subjective insights, he tries to hide his sin and shame by putting in some quotes.

What is more important for my purpose, however, is that Tinbergen ends up his *Study of Instinct* with six pages devoted to " The Ethological Study of Man ", and this brings me back to where I began.

SOLIPSISM

When objectivism, as applied to animals, is changed from a method into a dogma, it leads its adherents to write such words as " hunger ", " fear " and " pain " between quotes. They shrink from putting, even *prima facie*, the natural intuitive interpretation on the behaviour of animals. But if they should resolve to be consistent they would have to apply this principle to human beings also, and then it would lead them straight into solipsism. Let us see, then, what light solipsism can throw on our subject.

I feel a strong conviction that you exist, and you feel a strong conviction that I exist. So strong are such convictions that many people are incapable of questioning them. But is their strength due to strength of the evidence, or merely to the force of habit ? The answer is that it cannot be due to the strength of the evidence, for a careful scrutiny of the latter will show it to be very thin. Its thinness can most easily be seen by remembering that in a vivid dream it is possible to see and argue with people who, as one now admits, were not really there. It is even possible to dream that one has just awakened from a dream, or that one is arguing with somebody as to whether one is dreaming or not. Most philosophers shirk this difficult issue, but it is interesting to see how John Locke (1689*b*) dealt with it. " If any one will be so sceptical," he says " if he pleases he may dream that I make him this answer, that the certainty of things existing *in rerum natura*, when we have the testimony of our senses for it, is not only as great as our frame can attain to, but as our condition needs. For our faculties being suited not to . . . a perfect, clear, comprehensive knowledge of things free from all doubt and scruple but to the preservation of us, in whom they are . . . they serve our purpose well enough, if they will but give us notice of those things which are convenient or inconvenient for us." He then gives as an instance the burning of your hand if you put it in a flame. Thus, he resorts to what would now be called a positivist solution. But in this way he misses the point, which is that we shall feel very lonely and unhappy if we may not assume that the conscious minds of our friends really exist, as noumena that never become phenomena for us ; as entities which are real but not observable. Dogmatic objectivists must repudiate this assumption ; they must live as solitary solipsists in a lonely universe, if they are to be consistent. Ayer (1956) tried to solve the problem in some pages of diffuse semantics, but he lost it in a sea of words ; for the real difficulty here is not to analyse meaning but to validate belief. Herbert Dingle (1955) devoted to the same subject twenty-two pages which contain some sound common sense, but he also missed the point.

It seems to me that the only valid answer to solipsism, which will also point to a final justification of anthropomorphism, is one given by Bishop Butler (1736) in the parallel case of atheism. It amounts to this, that while in academic speculation our conclusions may, if we choose, be determined by probability alone, in practical affairs they must be determined not by simple probability but by something like what mathematicians call " expectation ". Thus the answer to solipsism is this, that if I assume other people to exist when in fact they do not, as certainly happens in dreams, no harm will be done ; whereas if, in the converse case, I assume that people do not exist when in fact they do, great harm will result. The same principle applies to anthropomorphism. If I assume that animals have subjective feelings of pain, fear, hunger and the like, and if I am mistaken in doing so, no harm will have been done ; but if I assume the contrary when in fact animals do have such feelings, then I open the way to the unlimited cruelties for which we execrate the Port-Royalists. This is the conclusive defence of anthropomorphism : animals must have the benefit of the doubt, if indeed there be any doubt. And if I am asked how low down in the animal world such feelings must be imputed, the answer is plain : any organism which is capable of learning by rewards and punishments must be assumed to be capable of feeling pleasure and pain in the absence of proof to the contrary, and the only conditionable animals of which the contrary can be proved are Grey Walter's electronic tortoises.

CONCLUSION

I have tried to show that the objections to anthropomorphism must lead the objector to become an anti-vivisectionist or a solipsist or both, if he is to be consistent, and that on the other hand an anthropomorphic understanding of animal behaviour, provided it be judiciously controlled by reference to observable facts, is justified by our intuitive interpretation ; by the experience of those who have to deal with animals ; by the theory of evolution ; by the behaviour of animals in response to injuries ; by the flexibility of instinctive behaviour ; by the rudiments of language which many species possess ; by electroencephalography ; by the existence of psychosomatic maladies in animals ; and by the ability of animals to learn or be conditioned. If, in spite of these considerations, anybody insists on using quotes when he refers to mental states in animals he has a right to do so, of course, but only on condition of being consistent ; so that if he should have occasion to tell a friend that his wife still loves him, he must put the word " loves " between quotes as well.

REFERENCES

Ayer, A. J.: 1956. "The Problem of Knowledge," pp. 36, 81-84, 243-254 (Macmillan).

Baker, John R.: 1948. "The Scientific Basis of Kindness to Animals" (UFAW).

Butler, Bp. Joseph: 1736. "The Analogy of Religion," pp. 188, 189 (Everyman).

Comte, Auguste: 1838. "Cours de Philosophie Positive," tome 3, (here translated). (a) 45e leçon passim. (b) p. 834 (c) p. 774. (d) p. 783.

Coulson, C. A.: 1955. "Science and Christian Belief," pp. 66, 67. (O.U.P.)

Crombie, A. C.: 1952. "Augustine to Galileo," p. 233 (Falcon).

Darwin, Charles R.: 1889*a*. "The expression of the Emotions in Man and Animals," pp. 13-15 (2nd edition).

— 1889*b*. ibid. pp. 317, 323, plate III (6), plate VII (2).

Descartes, René. 1637. "Discours de la Méthode," end of part 5. Cf. *Oeuvres Complètes*, Vol. IX, p. 426 ; Vol. X, p. 204 (Paris, 1825).

— 1649. "Lettre à M. Morus, 15 mars, 1649." *Oeuvres Complètes*, vol. X, p. 208. Cf. pp. 207 and 224-227. (Paris, 1925).

Dingle, H.: 1955. "Solipsism and Related Matters." *Mind*, Vol, LXVI, p. 433.

Grindley, G. C.: 1933. "The Sense of Pain in Animals" (UFAW).

Hume, David: 1738. "A Treatise of Human Nature." Bk. 1, pt. III. sec. xvi.

Jeffreys, Sir H.: 1957. "Scientific Inference," pp. 171, 188 (O.U.P.).

Lack, David: 1953. "The Life of the Robin," pp. 156, 165, 167 (Penguin).

Locke, John: 1689*a* "An Essay Concerning Human Understanding", Bk. 2, chap. 10, sec. 10 (19th edition).

— 1689*b* ibid. Bk. 4, chap. ii, sec. 8.

Lorenz, Konrad Z.: 1950. (trs. 1952): "King Solomon's Ring."

Mill, J. S.: 1865. "Auguste Comte and Positivism," p. 6.

Morell, Thomas: 1793. Notes in the nineteenth edition of Locke's "Essay concerning Human Understanding", *ad loc.* on p. 34. (G. Sael, London 1794).

Morgan, C. Lloyd: 1894. "Introduction to Comparative Psychology," p. 53.

Pumphrey, R. J.: 1952. "Ethology comes of Age," *Advancement of Science*, Vol. 8, pp. 376, 377.

Raven, Charles E.: 1942. "John Ray, Naturalist," pp. 374, 375.
— 1951. Gifford Lectures, p. 109.
Ray, John: 1692. "The Wisdom of God in Creation," pp. 42-44.
Romanes, G. J. 1882. "Animal Intelligence," pp. 2-5 (Kegan Paul).
— 1883. "Mental Evolution in Animals," chap. 1 (Kegan Paul).
Sainte-Beuve, C. A.: 1878. "Port Royal," Bk. 2, chap. 16 and note 21 thereon.
Sextus Empiricus: "Pyrrhonis Eliensis Vita, ex Diogene Laertio."
Spearman, C.: 1927. "The Abilities of Man," p. 167.
Tinbergen, N.: 1951. "The Study of Instinct," p. 3. (O.U.P.).
Tinbergen, N.: 1953. "Social Behaviour in Animals," p. 7 (Methuen).
Watson, J. B.: 1925. "Behaviourism," p. 6 (Kegan Paul).
Wells, H. G., Huxley, J. and Wells G. P.: 1937. "How Animals Behave," p. 200. (Cassell).

How to Befriend Laboratory Animals

First published in 1949. Last revised in 1962.

HISTORY OF THE PROBLEM

IN 1863 it became known that veterinary students in France were being required to perform, as part of their training, operations on live horses without anaesthetics; a series of operations on each horse. A memorial signed by 500 British veterinary surgeons was taken to Alfort by a Mr James Cowie, and the practice was subsequently discontinued. The shock it had given to British public opinion led, however, to discussion of experiments on animals as practised in this country; lay opinion was aroused by the writings of Miss Frances Power Cobbe, and on the scientific side the British Association studied the subject on the initiative of Charles Darwin, Lyon Playfair and other humane scientists. The report of a Royal Commission on Vivisection, published in 1876, resulted in the Cruelty to Animals Act (1876), which is still in force and protects research workers from vexatious litigation but provides important safeguards for animals. A second Royal Commission reported in 1912 and made recommendations as to the administration of the Act; most of these have been adopted.

At first the anti-vivisectionists, led by Miss Cobbe, co-operated with humane scientists, but when the Act was passed they were very dissatisfied with it. They were, in fact, in a very difficult position. They were filled with horror by the Alfort incident and some other occurrences that had come to light; they had a notion that they had been betrayed in some way by their scientific friends; without having had any scientific training themselves, they had to deal with highly technical subject-matter; and they wished to appeal to the lay public, which was almost as unwilling then as now to understand anything that is at all difficult or complicated, or to take notice of anything that is not said in a sensational way. In these difficult circumstances, groping in a half-light and tormented by an urgent sense of horror, they adopted methods that have characterized much anti-vivisection propaganda ever since. They simplified their policy by demanding the total abolition of all experiments on animals without discrimination, simplified their polemic by asserting that experiments on animals cannot usefully advance medical knowledge, and simplified their indictment by alleging that all experiments are cruel and all experimenters heartless. It is easy to understand how this mode of propaganda arose, but equally easy to understand the intense resentment aroused by it in the minds of scientists. That resentment discouraged many from making the stand they might otherwise have made against cruel experments. They became afraid of seeming to countenance indiscriminate anti-vivisection propaganda, and preferred, like Charles Darwin after the schism, to "bear their share of the abuse poured in so atrocious a manner on all physiologists". On the other hand, whatever their shortcomings, the anti-vivisectionists have kept us aware of the problem. However much we may disagree with their policy, that fact entitles them to our gratitude.

FAILURE OF A PROPOSED SOLUTION

The abolitionist policy, that is to say the demand for the abolition of all experiments on animals without discrimination, has not been successful. Since it was promulgated the number of experiments performed annually has increased rapidly; in 1883 there were 311 experiments, and in 1961 nearly 4 million. No anti-vivisection Bill has reached second reading in either House since 1881. In 1929 when an anti-vivisectionist, Mr. J. R. Clynes, was Home Secretary, Sir Robert Gower asked him in the House of Commons to introduce legislation prohibiting experiments on dogs; he refused on the ground of advice tendered by the Government's professional advisers. If the abolition (as distinct from regulation) of experiments on animals ever became probable, the Medical Research Council, the Agricultural Research Council, the Royal Society, the B.M.A., the N.V.M.A., the Royal Colleges, the Ministries of Health, Agriculture and Food, the Home Office, the Colonial Office and the Defence Services, backed by scientific and medical opinion, would ensure its being averted.

On the other hand it is illegal to perform an experiment on a living animal for the purpose of attaining manual skill, but this enactment is apt to work out rather badly; its effect is that all veterinary surgeons and some physiologists, instead of practising upon animals that are destined to be killed before they return to consciousness after being anæsthetized, must acquire manual skill at the expense of animals that are intended for recovery and will suffer, therefore, if an operation should be bungled. Again, it is true that the Dogs Act, 1906, prohibits the use of stray dogs for vivisection, but the effect of that is that dogs bought from dealers have to be used instead, and the resulting high prices tempt thieves to deprive dogs of their homes. It is true also that according to law a horse must not be used if any other animal will do, and a cat or dog must not be used for experiments without anæsthetics if any other animal will do. Horses, cats and dogs are such favourite animals that they are likely to be treated with more careful consideration than other species in the laboratory, and the effect of the law is to substitute for them animals which are equally sensitive to suffering but less likely to receive specially considerate treatment.

These are, I think, the only instances in which the abolitionist policy has been realized, and although they give satisfaction to pet-owners they are disadvantageous to animals.

AN ANALOGY

In the nature of the case the abolitionist policy was bound to fail; it is as if I were to demand the abolition of the motor-car in view of the fact that in 1960 no fewer than 6,970 persons were killed and 340,581 injured in road accidents in Britain, often with cruel mutilation, and that

the same sort of thing happens year after year. In support of my thesis I might describe in gruesome detail some of the injuries inflicted on attractive young children. I might allege that motor-cars are unnecessary, and our ancestors were happier without them, as the inhabitants of Sark still are; that is possible to cross London by tube a good deal faster than by taxi; that cars break down, and freeze up, and get ditched, and catch fire, and skid into shop windows, and menace the morals of innocent young women. As to the effectiveness of mere palliatives, I could point to the pedestrian crossings that most pedestrians ignore, the traffic policemen diverted from other duties, the safety-first weeks and pathetic posters, and the ghastly toll of life and limb that accrues to the motorist every week in spite of all. I could allege that every driver is a Toad of Toad Hall at heart and every road-house bar a Sybaris seething with tipsy road-hogs. I could paint a picture of the highways of this country as a gladiatorial arena filled with sadistic speed-maniacs in pursuit of agile and terrified pedestrians, and strewn with the corpses of honest cyclists who have been butchered to make an English bank holiday.

I could put out a very strong propaganda for the abolition of the motor-car, and arouse strong feelings against motorists, but my prospect of success would be pretty much the same as the prospect of abolishing all experiments on animals. Parliament, guided by the Government's technical advisers, would prefer to fall back on such expedients as imposing a driving test, invoking the good will and decent feeling of the motoring fraternity, using traffic lights and traffic policemen and zebra crossings, and in the last resort taking legal action against outstanding offenders.

WHAT IS THE USE OF EXPERIMENTS ON ANIMALS?

Medicine is partly empirical and partly scientific. Migraine is often treated empirically; the doctor tries various palliatives that have been known to give some relief, and if he finds one that suits his patient a little better than others he may not know why it should do so. Experiments on animals are of little use to empirical medicine; it is useless to try such things on the dog because the dog's reactions may be quite different from those of a human patient.

Scientific medicine, on the other hand, is based on an understanding of the way the body works, and on the natural history of bacteria and protozoa and helminths.

All mammals have blood, lungs, stomachs and nerves, and the differences as well as the resemblances between species make possible the comparative method which is a powerful organon of biological research. By means of controlled experiments a general understanding of the laws that describe organic life can be built up, and in the light of that understanding medical treatments can be devised. The difference between empirical and scientific medicine is like the difference between an amateur changing one or two parts in a faulty wireless-set in the hope that it will come right, and a wireless mechanic who knows the circuit going over it with voltmeter and oscilloscope. Now medicine, in so far as it has become scientific, depends on physiology, pathology and pharmacology, and those sciences could not have progressed so far as they have without experiments on animals.

That is not to say, by any means, that every experiment leads to useful knowledge. Some very cruel experiments have been ill conceived, badly designed, unnecessary, and wastefully carried out. Nevertheless, to admit all this is very different from saying that, by and large, experiments on animals have not yielded useful knowledge; that would be not only untrue but fantastic, like saying that the earth is flat or that rheumatism is caused by witchcraft.

THE RISK TO WHICH LABORATORY ANIMALS ARE EXPOSED

Pain is a sensation, like touch, sight and hearing, and is something totally different from intelligence and self-consciousness. In animals and men alike there are pain-nerve fibres, no less than optic nerves and nerve fibres for sensing heat, cold, sound and muscular tension. Sensation is a more primitive mode of consciousness than thought or introspection, and the senses of animals often have a lower threshold than those of men; as for instance, has smell in the dog, night vision in the cat, and touch in the pig's snout. An authoritative treatment of the sensibility of animals to pain will be found in the UFAW monograph *The Scientific Basis of Kindness to Animals*, by Dr John R. Baker, F.R.S., and in the *International Symposium on the Assessment of Pain in Man and Animals*, published by UFAW.

It is a tragic error to suppose that because animals are less intelligent than men they therefore feel less pain. A young baby is less intelligent than a kitten of the same age, but not less sensitive. Emotions such as fear are more primitive than thought or reason, and animals appear to have emotions as strong as, if less complicated than, those of human beings.

Laboratory animals are exposed to nine risks :—

(1) Certain animals, the horse, cat and dog, are more generally understood and consequently more popular than other species. In a foreign country experiments done on dogs under anæsthetics have been repeated on less popular animals without anæsthetics. There is a danger that animals like rats, which happen to be unpopular but are just as sensitive as are the customary pets, may be sacrificed

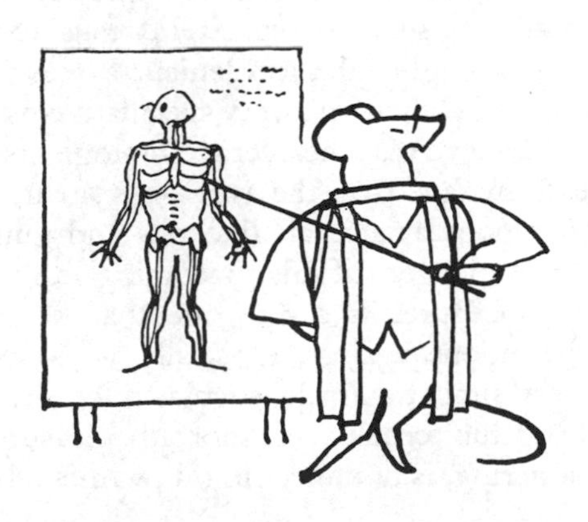

in deference to ill-informed popular sentiment.

(2) When a research worker has to adjudicate between the claims of his own research and those of the material he is using in it, he is likely to be biased in favour of the former.

(3) A doctor's job is to care for human life and limb, and it is an extremely exacting one, so much so that there is a risk of his under-valuing considerations other than those which concern his profession. When, therefore, the claims of laboratory animals have to be balanced against those of human medicine a medical scientist might be biased. A zoologist or a veterinary surgeon, who on one hand is a human being himself and on the other is a student of animals, might find it easier to take an unbiased view.

(4) A person who feels it his duty to inflict some degree of suffering is in danger of having his sensibility blunted in the course of time in respect of what at first was repugnant to him, and the more sensitive he is by temperament the more he tends to shield his feelings by repression. This hardening effect seems to be largely specific; a person becomes callous about things he is accustomed to do, but not necessarily about anything else. The old rule that butchers should not serve on juries in trials for murder, for fear that they might take too lenient a view of mortiferous practices, was based on a very speculative psychology.

In general, I believe that the average biologist is probably more, not less, humane than the average layman, especially in respect of unpopular animals like rats and guinea-pigs; nevertheless the danger of his becoming callous about experimental procedures is a real one that he must take deliberate precautions against, especially when very large numbers of very similar animals, such as mice, are in use.

(5) Psychopathic cruelty, a morbid pleasure in the infliction of suffering, is not uncommon; witness the crowds

who used to flock to see a hanging. Moreover it can coexist with acute intellectual gifts, as it did in the Marquis de Sade himself, and there is therefore a possibility that sufferers from this complaint might be attracted into research of the more drastic kind, such as studies of shock. I do not know of any such case in Britain, but there are clear indications of sadism in some foreign experiments. In certain excessively cruel foreign work on rats the number of animals used was far in excess of what was requisite for the purpose in view, and in some cruel psycho-physiological experiments on dogs the experimenter spent "from one to seven hours almost daily" watching the animals during seven months in one case and twelve months in another. Both excessive sample-size and gloating are prima-facie indications of sadism.

(6) The psychological requirements of the animals used have not hitherto been sufficiently studied. UFAW is sponsoring a research on this subject.

(7) In most countries there is nothing to prevent scientifically incompetent persons from experimenting on animals. In the United States children are allowed to send up live mice in rockets in the belief that they are being scientific, and have been encouraged to take part in quasi-scientific activities which include toxicity tests, intraperitonal transfer of cancer cells, the effect of heat on cancer growth, and even perfusion experiments. "Surgical procedures are especially thrilling to pupils", who have been allowed to "take mice and other animals home to use in experiments conducted in their own rooms and basements." They have been encouraged to cultivate what is called an "impersonal and objective", *i.e.*, a non-sympathetic, attitude towards the animals. Commercial firms have offered prizes and facilities for children's experiments, and these are popularized in the annual Science Fair.

(8) Incorrect or callous husbandry and lack of trained animal technicians are serious problems in some countries.

(9) Beginners need training in responsibility for the animals they use.

HOW MUCH DO LABORATORY ANIMALS SUFFER IN BRITAIN?

A clear distinction must be drawn between laboratories in Britain and those in other countries. A number of the latter have laws regulating experiments on animals, but most have either no such law or inadequate machinery for implementing the law.

The system existing in this country is very British. It is archaic and illogical and yet it has worked very well, though its task has somewhat outgrown its resources. A person who knew no more about it than what he could learn from the wording of the Act of 1876 might suppose it to be mere eye-wash, but anybody who thinks it really is mere eye-wash should compare British with American experimentation and then think again. A kind of traditional gentlemen's agreement has grown up between scientists and the Home Office, and though one may wish it to be improved in certain ways it affords a unique degree of protection to laboratory animals.

The law, which is rather complicated, is summarized in UFAW's pamphlet *Suggestions for the Legal Protection of Laboratory Animals Overseas*, and only some of its practical effects will be mentioned here. Every experimenter must be licensed by the Home Office, and licences are by no means to be had for the asking. For experiments on conscious animals a certificate as well as a licence is required. Permission is never given to perform an operation more serious than simple inoculation or superficial venesection on any vertebrate, be it dog, rat or frog, without anæsthesia. Most of the operated animals are allowed to recover consciousness and their condition thereafter depends on the skill and care with which they are nursed as well as on the nature of the operation.

Most experiments do not involve any surgical operation and in these cases anæsthetics are usually dispensed with. Although in strict law the Act applies only to " experiments calculated to cause pain ", by a gentlemen's agreement all experiments which are calculated to interfere with the normal health or comfort of an animal are dealt with under it. Some involve no suffering at all ; for instance animals may be infected and then killed before their malady becomes painful, or a sample of blood or a vaginal smear may be taken after administration of a hormone. Other experiments involve some discomfort, such as moderate fever or the formation of a tumour which does not reach a painful stage. But there are others again, even in Britain, which are more serious and involve disagreeable symptoms ; these are very much less numerous but they exist.

All experiments on conscious animals are subject to the Pain Rule, which requires *inter alia* that any animal found to be suffering severe pain which is likely to endure must forthwith be painlessly killed ; and this must be done whether the main result of the experiment has been achieved or not.

The Home Office has a staff of six inspectors who visit each laboratory two or three times a year on the average. The second Royal Commission recommended on the basis of the 1905 figures that the number of inspectors should be doubled, and the number of laboratories has increased so much since then that there would have to be a great many more inspectors if the recommended scale were to be reached ; on the other hand the quality of the inspectors matters more than their number. You cannot, of course, deal with responsible scientists on a detective-and-suspected-criminal basis, and although the inspectors have disciplinary powers in reserve their primary function is to advise licensees who must be presumed to be humane, and in fact appear to be so with only rare exceptions. The inspectors ought, however, to be able to consider all types of experiment minutely, bring precedents under periodical review, go fully into the humanitarian and technical qualifications of applicants for licences, and carry out several other duties, but their task may well have outgrown their numbers.

Apart from experimental procedures there is the important question of the housing, care and comfort of the animals. This is a serious matter, and one in which a great deal can be done for their benefit. A great deal of attention is in fact being paid to it as a result of the work of the Laboratory Animals Centre and Animal Technicians' Association and the publication of *The UFAW Handbook*, and in this connection let us bear in mind that the greater the care and affection lavished on the animals in the animal room, the greater the consideration that will be shown to them in the laboratory. Psychological attitudes are infectious.

HOW MUCH DO LABORATORY ANIMALS SUFFER IN FOREIGN LABORATORIES?

Descartes's theory of vortices survived Sir Isaac Newton by some 50 years on the continent, but his even more preposterous theory of the robot animal has survived Charles Darwin longer yet, and still colours the Latin attitude on such subjects. The late Professor E. H. Starling once assured me that much French and Italian physiology had been vitiated by failure to eliminate the effects of pain, and when I mentioned this remark to Sir Joseph Barcroft it brought out the comment that he once had to quit a French laboratory because the cruelty going on there was going to make him vomit. These two were physiologists of world-wide reputation, not likely to be sentimental or squeamish.

In most foreign countries there is no law for the prevention of cruelty to animals, in laboratories or anywhere else, and all too little mercy of any kind for man or beast.

Responsible American research workers have a voluntary ethical code and presumably many conform to it, though here again the less popular animals seem to come off worst. The code cannot be enforced by law, and instances of prolonged torture of a wholly unjustifiable nature occur. While there are laboratories, such as those of Parke Davis, where the husbandry is excellent, there are others in which it is backward and there is insufficient regard for the comfort or feelings of the animals. Animals are wastefully used, funds being poured out like water for experimentation by both experts and dabblers. British scientists who have worked in the United States have often been shocked by what they have seen. The Animal Welfare Institute of New York is advocating legislation based, *mutatis mutandis*, on British experience, and carries on an educational campaign in favour of good husbandry and handling of laboratory animals, and of sane methods of teaching biology in schools, but it faces bitter and very unscrupulous opposition from vested interests.

Again, the British Commonwealth as a whole lags far behind the mother country.

On the other hand the International Committee on Laboratory Animals, recently formed with the aid of UNESCO, is in good hands and is likely to do good work for animal husbandry.

ETHICS: (1) SUFFERING NOT ARISING OUT OF TECHNICAL REQUIREMENTS

Much unnecessary suffering can be prevented if sufficient trouble is taken by a competent staff, and in the best laboratories this is done. In Britain there is a strong movement afoot for bringing the care of animals in all laboratories up to the standard of the best. The techniques of anæsthesia (with precautions against premature recovery) and euthanasia; operative skill and post-operative care, including the use of drugs to ease post-operative pain; the selection and training of Laboratory Technicians and Animal Technicians; housing, feeding and watering; hygiene, bedding, light, shelter and ventilation; company, play and exercise; gentle, skilful and frequent handling; a considerate recognition of the fact that sentient creatures are not test tubes nor machines; all such things can make the difference between misery and a maximum of comfort. There can be no legitimate difference of opinion about their importance.

Here is another matter that does not admit of any legitimate difference of opinion. When an experiment involves any degree of suffering it is morally wrong to use animals wastefully. For instance, homogeneous stocks of rats, mice and hamsters show far less variability than animals bought from a small dealer, so that fewer of them are needed for a given degree of precision. Further, the factorial planning of experiments with a view to an analysis of variance often reduces greatly the number of observations required; sequential sampling may possibly lead to a further reduction; and recently it has been found that certain specific factors may affect variance. Again, economy in sampling can be promoted (*i.e.*, waste can be avoided) by choosing the right strain of the right species for each particular research. Finally, a preliminary search of the literature will obviate the waste of animals on unnecessary repetition of experiments. In the United States the fantastic increase in amateurish experimentation is making such a search difficult or impossible. It will be agreed by everybody that there is a moral obligation to take all such precautions in order to prevent the useless multiplication of suffering.

ETHICS: (2) MAN AND BEAST

The matters discussed in the preceding paragraph do not admit of any reasonable difference of opinion, but we now come to a more difficult question. When suffering cannot be avoided if a given experiment is to be carried out, where is the line to be drawn? We are faced with the perplexing problem of choosing the lesser of two evils and saying how far the end can justify the means. Clearly it cannot do so without limit. The Nazis by sacrificing a few prisoners in controlled experiments sought to save the lives of thousands of Germans, but most people hold that they were wrong in doing so, however valuable the results they hoped to achieve. The utilitarian principle breaks down if you push it too far.

On what basis are the interests of animals to be weighed against those of man? Extremists on one side and the other feel no difficulty. Anti-vivisectionists say that no experiments ought to be done on animals at all, but take this case: One method of assaying narcotics consists in giving a minute dose to a rat who has learned to run through a maze for his dinner, and noting how far the narcotic slows him down by comparison with his established handicap. Can anybody call that cruel? At the other extreme we have fanatics who say that mere animals must always be sacrificed to human interests; that any outrage, however horrible, may be perpetrated against an animal if the human race stands to gain by it. But take this other extreme case. Veal used to be made by hanging up a calf by the hind feet, bleeding it slowly, and maltreating it meanwhile to make it struggle. The extra white veal might tickle a gourmet's appetite, but is this worth that?

Setting aside both extremes a moderate person has to draw the line somewhere between them, and the question is, where?

We must distinguish clearly between killing and hurting. To kill an animal is a far less serious matter than to kill a human being. For one thing only a small percentage of animals can reach maturity in any case, for otherwise there would soon be no room left in the world; hence we kill unwanted puppies and kittens. There is no harm in killing animals provided it be done painlessly. For instance, a method of testing milk for tubercular infection consists in injecting a preparation into guinea-pigs, killing the animals painlessly before the disease has reached a stage where it can cause any suffering, and then making a post-mortem examination. Nobody can reasonably object to that.

But when it comes to hurting, as distinct from killing, is there any reason why hurting an animal should be less objectionable than hurting a man with the same intensity? We must not allow this issue to be confused by the side-issue of social repercussions. If you hurt a man, you also cause distress to his relatives and friends and you may incapacitate him from carrying out his social duties; these considerations do not arise in the case of an animal and they have to be taken into account, but they are irrelevant to the principle of our question and must not be allowed to confuse it. Apart from these social repercussions, is it more objectionable to hurt a man than an animal? I should say definitely not, and if anybody thinks that it is, it is pertinent to put the question "Why?" Is it on the ground that animals have no souls, that is to say no future life? But if it is true that they have none, there is all the more reason for making happy the only life they are to have. Is it because animals are less intelligent than human beings? If so, children ought to be hurt in preference to adults, who are more intelligent than they. Is it because a human personality is more valuable than that of an animal? Its value is irrelevant to hurting, though not to killing: to cause pain to a mentally sick person is not preferable to causing pain to a hale-minded and therefore more valuable one. Is it that animals are less lovable than human beings? I kept pet rats as a boy and lived among wild ones in the trenches, and found it as easy to love them as I have found it difficult to love Nero and Louis XIV and Himmler, and some further east that I could mention, and for that matter some nearer home as well. If I am asked "What about your own relations and friends?" I reply that I might be tempted to commit perjury to save one of them from the gallows, but that would not make perjury right. I know of no reason except prejudice for preferring that an animal rather than a random human being should suffer a given amount of pain, provided always that individual suffering be distinguished from its social repercussions and from the risk of death.

ETHICS: (3) SUFFERING ARISING OUT OF TECHNICAL REQUIREMENTS

When an experiment cannot be carried out without the infliction of suffering, in what cases can the end justify the means? To answer this question, considerations which are quantitive and yet cannot be measured have to be balanced against one another. That is a difficult task, and gives rise to differences of opinion between humane and reasonable persons, but it is what any jury has to do when there is a conflict of evidence. The difficulty of the task is a reason for devoting care and thought to it, but not for shirking it or taking refuge in extreme views or short cuts.

Experiments calculated to cause pain are permitted in Britain (subject, however, to the Pain Rule) for the advancement of either (i) "physiological," *i.e.*, biological, knowledge,

or knowledge useful for (ii) saving or prolonging life o
(iii) alleviating suffering.

(i) As regards fundamental or pure biology: the mere satisfaction of scientific curiosity cannot justify the infliction of serious suffering. One can never tell, of course, whether any given item of fundamental knowledge may turn out to be useful for the prevention or relief of pain, but only direct relevance to some medical application for that purpose can justify experiments involving appreciable suffering.

(ii) The mere prolongation of bodily life is not regarded as of over-riding importance in Christian ethics, nor by those who, in war or otherwise, sacrifice their lives for an ideal. I do not see how seriously painful experiments can be justified for such a purpose, quite apart from the terrible prospect of over-population which already threatens civilization with disaster. Personally, I would most certainly not wish my own and others' lives to be prolonged at the cost of such tortures as have been inflicted in New-World experiments on shock, and I believe that anyone who thinks the thing out must come to the same conclusion.

(iii) When knowledge that will prevent a large amount of suffering can be obtained by inflicting a relatively small amount, then many will hold that the end may justify the means, though again it is a question of degree. It must be remembered, however, that the value of a research is always problematical till the work has been done. At the very least the objective must be clear and fully adequate, the probability of success high. Every precaution, including economy in sampling, must be taken to reduce suffering to a minimum, and there must certainly be an upper limit to what may be done. In no case would it be justifiable to suspend the Pain Rule, for instance. The second Royal Commission, five out of whose seven members were medical, laid down that rule absolutely and without qualification.

ETHICS : (4) THE CRITERION

If it be assumed that the infliction of appreciable suffering may be justifiable in exceptional circumstances, when there is a high probability of relieving much greater suffering by means of the knowledge to be gained, a criterion is needed for enabling the decision for or against to be fairly taken in any particular case. A suitable criterion for use by the person who is to make the decision can be reached in the following way. He must begin by putting himself in imagination in the place of the animal. This can be done : every successful animal-trainer has to do it ; animal-psychologists do it, some deliberately like Bierens de Haan and some while pretending not to, like the behaviourists. The experimenter or inspector must, then, put himself in imagination in the place of the animal. He must ignore social repercussions and the risk of death (which alone can justify him in choosing a victim other than himself) and focus his attention on the individual experience of pain or fear involved ; and he must now ask himself "Should I myself be willing to endure that degree of suffering in order to attain the object in view ?". If not, his decision must be negative, and no experimenter has a right to do to an animal what he would not be willing to have done to himself, but for any social repercussions or risk of death that might be entailed.

POLICY

In view of the technical nature of the problems that arise, the last word in the regulation of experimental biology must always rest with scientists. The wise and realistic policy for animal-protectionists is therefore to enlist the sympathy and co-operation of the scientific world, and this is not difficult. Experience has shown that scientists, including medical men and vivisectors, will co-operate cordially with other animal-lovers if approached in a reasonable way, and many of them are deeply concerned about the more painful sort of experiments, as well as about the proper care of the animals. What is needed is the frank discussion of the very difficult problems that arise, and that is impossible in the cut and thrust of controversy and mutual recrimination.

A vivisector who has held a Home Office licence for a good many years wrote this to me with regard to certain experiments done in the New World: "It seems to me that the animal welfare movement is a *farce* if no objection is made to experiments such as these. I did not know of their existence till you put me on to them. If that is science, I should rather clean out a lavatory for a living than call myself a scientist—far rather, for that would be a useful and honourable job." I quote this private letter, written in the heat of the moment and not for publication, in order to show our anti-vivisectionist friends that it is possible for an experienced vivisector, a keen scientist, to feel as strongly as they do themselves about experiments which are cruel. In May, 1957, about 120 research scientists attended a symposium organized by UFAW on "Humane Technique in the Laboratory". Most of the papers were published under the auspices of the Medical Research Council, and two were published in full in *The Lancet*.

The behaviour of individuals is governed largely by the standards prevailing in the circles to which they belong. The surest way of befriending laboratory animals is therefore to cultivate inside the scientific world a public opinion actively concerned with their interests. The lead must be taken by persons with scientific qualifications, but laymen can give sympathetic support if they approach scientists in a sane and knowledgeable way.

Since science is international it offers perhaps the most

promising channel for conveying British ideas of the right treatment of animals, not only inside but also outside laboratories, to less advanced countries. Britain was notorious for cruelty a century and a half ago but has made such advances since then that most other nations are backward by comparison. With this in mind UFAW has published *Suggestions for the Legal Protection of Laboratory Animals Overseas*.

IMPROVEMENTS

Britain has reason to be proud of the way she has handled the problem of animal experimentation, but in this as in all other human undertakings constant progress is possible. A few feasible improvements will be suggested.

An increase in the number of Home Office inspectors is desirable, not for police purposes but in order that their task of promoting a uniform standard of humaneness may be more easily carried out, with maximum efficiency and plenty of time to think. The Home Office inspectorate could advantageously include one or more veterinary surgeons who could advise on animal management. It is desirable that existing precedents should be continuously reviewed in relation to the Pain Rule, and that zoologists, veterinary surgeons and judicious animal-protectionists should be consulted regularly.

Special consideration should be given to the unpopular species, such as rats and mice, which are less sure of being treated with care than are dogs and cats.

Every encouragement is deserved by what is being done to improve the training, status and selection of laboratory technicians and animal technicians, and the status of the latter might be improved by generalizing the provision of a ladder of promotion affording a good career to entrants of the right type. Some system of awards for distinguished service in the animal room might be worthy of consideration. Similarly, the work of the Laboratory Animals Centre for raising the standards of accommodation and husbandry deserves the applause of animal-lovers.

Expenditure on animal rooms is needed on a much more generous scale than has hitherto prevailed. On this important point many scientists agree ; it is for the lay governors of hospitals and research institutions to give better priority to improving accommodation. But scientists might sometimes allocate a better share of their research grants to cages and similar equipment. Some of the commercial laboratories set a good example.

It is very desirable that scientific papers should specify, with all necessary detail, the technique of any anæsthesia that may be used, and any other special precautions against causing suffering. This is particularly valuable for the edification of beginners in research and of foreign readers.

The psychology of laboratory animals has not hitherto been sufficiently studied in relation to their mental comfort ; the conditions under which they are normally kept are not necessarily optimal from their point of view, and the factors which matter to them need to be identified. Continuous study of the techniques of anæsthesia and euthanasia also is called for. Researches are being carried out at Birmingham and in the Royal Veterinary College respectively, under the auspices of UFAW, with these two purposes in view.

In the assay of therapeutic substances a choice of alternative techniques is often available. Those are to be preferred which do not involve discomfort or suffering, and hard work is needed in devising humane techniques in cases in which none is available at present. Resistance to change of factory routine is natural but to be avoided. Precision which is swamped by variance in the response of patients to whom a drug will be administered is undesirable as entailing a wasteful use of animals.

ENVOI

To sum up, it seems that anti-vivisectionists took a wrong turning after the passage of the Cruelty to Animals Act, 1876, and animal-protectionists ought now to go back to their original policy of co-operating with humane scientists in order to obtain the maximum of practical protection for laboratory animals. The considerable success of the system administered by the Home Office points out the direction in which effective advantages can be obtained. Ethical abstractions, however sincerely believed in, do not really help at all. If we are to befriend laboratory animals effectively we must deal with the facts as they actually are.

In only a few foreign countries is there any system for keeping experiments on animals in the hands of those who are scientifically competent to do them, and for setting a limit to the suffering which may be caused ; if British anti-vivisectionists have failed of their avowed aim, their fellows overseas have for the most part nothing at all to show for all their exuberance. It is to humane scientists, and especially to humane biologists, that we must look for leadership and effective action, but laymen can help if they will understand the difficult parts of the problem and approach the scientific world in an informed way. Above all it is necessary to create an atmosphere of sanity and objectiveness and intellectual honesty in which the very difficult ethical problems that arise can be discussed discriminately, without the excitableness and invective that the subject has come to be associated with in many minds.

The Strategy and Tactics of Experimentation

Read at the UFAW symposium on Humane Technique in the Laboratory, May 8th, 1957, and printed in *The Lancet*, Nov. 23rd, 1957.

STRATEGY

LLOYD MORGAN put a fox-terrier into a yard enclosed by a picket fence, and the dog had to find an answer to the question : " How do I get out of here ? " His method was to push his nose between the pickets in one space after another at random, and eventually he happened to lift the latch. " This line of attack is called ' trial and error,' turn out to be false leads or errors " (Woodworth and Marquis 1949). It may also be called " empirical ", in contrast with the following kind of animal behaviour, which entails what is called " insight ". One of Köhler's apes, having learnt to hook a banana into his cage with a long stick, had next to find an answer to the question : " How do I get hold of the banana with sticks that are too short to reach it ? ". In playing with two shorter sticks which could be fitted together he happened to join them. He at once ran to the front of his cage and tried the experiment of reaching for the banana with the compound stick, an experiment which gave a positive result (Köhler 1925).

Sometimes a research-worker whose mind runs to empirical or trial-and-error methods will get there in the end, but the born scientist uses insight; by a kind of intuition he hits upon a clue, uses the clue to ask himself the right questions, and then plans experiments to answer the questions, yes or no. But the trial-and-error empiricist, like Lloyd Morgan's terrier pushing his nose industriously into one space after another, does not begin to see his way to the goal until he has got there.

EMPIRICISM VERSUS INSIGHT IN PHARMACOLOGY

The point may be illustrated by reference to a review (*Nature* 1953) of a report written by two pharmacologists who had screened no fewer than 3,500 organic compounds for action against the tubercle bacillus. The reviewer commented as follows. The aim, he said, was to discover or invent "a chemical remedy which will eradicate, completely, virulent tubercle bacilli from human tissues without causing injury to the host cells. The natural approach for the chemist would be to design or select molecules likely to become involved in the bacillary metabolism". But the authors "believe that the only method that offers promise of early success is random screening, and they do not hesitate to claim that there should be no basis for selection of compounds other than availability. The sentence—'The random selection of compounds to be tested should be adhered to as strictly as possible'—has a kind of lunatic flavour about it".

The reviewer continued:

"Finding *the* chemotherapeutic agent for tuberculosis is a difficult problem, but it will scarcely help the search to throw reason to the winds. There are known nowadays a number of widely different types of molecule of synthetic and natural origin which exert a considerable suppressive effect on the tuberculous process in man and in animals. This knowledge provides a rational starting-point for structural tailoring which may lead to the production of substances of greater activity and lower toxicity. Alternatively, an intensive study of the mechanism by which any one of these substances interferes with the metabolism of the bacillus may provide an important clue which would guide further synthesis. This will be a difficult but perhaps not unrewarding investigation. If a greater proportion of the resources which have been used by commercial firms in screening potential anti-tubercular substances were diverted to the study of the chemistry and biochemistry of *Mycobacterium tuberculosis*, progress would be eventually more rapid."

The results of this vast application of the empirical method were meagre. Of eleven compounds shown to exert a repressive effect on tuberculosis, no fewer than ten had already previously been known to have this property, and the reviewer remarked: "Since the discovery of activity in these compounds by various workers was attained by following a lead, however slender, the heroic empiricism of the method of random selection does not inspire confidence."

We must make a distinction between this unjustifiably empirical method of searching for a substance having a prescribed property, on one hand, and on the other the testing of a totally new compound to find out what unknown properties it may have. Here a certain amount of trial and error is perhaps inevitable, though it can be kept within reasonable limits by means of hard thinking.

EMPIRICISM VERSUS INSIGHT IN GENETICS

Illustrations of the superiority of insighted research over empirical research might also be drawn from the history of genetics. At the beginning of this century the study of heredity was mainly in the hands of the Weldon school, who were industriously measuring biometrical correlations and collecting many data which seemed to get them nowhere in particular. In contrast to these empiricists, Charles Darwin had earlier put forward the explanatory hypothesis of pangenesis, which happened to be untrue but nevertheless stimulated thought. It was followed by Weismann's germ-plasm theory, which was a better approximation to the truth and served as a guide to experimentation. But genetics as we know it was born in 1900, when the rediscovery of Mendel's principles made possible an explanatory hypothesis or clue on the basis of which the decisive experimental work of the Bateson school could be planned. Prof. V. H. Blackman (1956) writes of an incident in the controversy that ensued:

"At a meeting of the Evolution Committee of the Royal Society, Weldon had read a paper on the sizes of the carapaces of a certain population of crabs. Bateson, who considered the results of no biological importance, when asked by the Chairman of the Committee to comment upon Weldon's contribution did so in a single devastating sentence: 'though all science might be measurement, all measurement was not necessarily science'."

Professor Blackman adds:

"I may say I have occasionally found it useful to bring this dictum before research students, who are sometimes inclined to believe that so long as they are measuring something they must be advancing science."

THINKING AND EXPERIMENTING

I have spoken as if there were a hard-and-fast line between empirical research, proceeding by trial and error, and insighted research which follows a clue. In fact, however, the difference is quantitative rather than qualitative; the ratio of the amount of thinking to the amount of experimenting that goes into an investigation varies from one worker to another. This ratio was small in the chemical screening project I have referred to, and in some of the biometricians' work on heredity. It was still smaller in a certain research which was undertaken with the object of finding out whether a rat can choose a balanced diet by instinct. Purified samples of the various constituents of a balanced diet, vitamin this and vitamin that, purified proteins, and purified carbohydrates, were offered to the poor animal in separate saucers, and it was invited to make its choice without any help from those blends of smells and tastes or other signs in relation to which the instinct, if it exists, must have been evolved in Nature. Needless to say, the unfortunate rat failed in his examination, just as we should fail if our papers were set in some foreign language which we did not understand, but the penalty for failure in his case was harsh.

At the other extreme, the ratio of thinking to experimenting was surely maximal when Archimedes, meditating in his bath, hit upon the hydrostatic principle; this was the occasion when he ran home through the streets with nothing on, shouting " Heureka " (which was not his wife's name, as a schoolgirl once supposed). The somewhat similar course followed by a modern crystallographer in inferring from a few X-ray diffraction patterns the structure of a complicated organic crystal, which may have as many as 100 different parameters, has been described thus by Sir Lawrence Bragg (1948):

" The analyst has before him the separate bits of a jigsaw puzzle and he has to fit them together . . . In making guesses at possible arrangements he draws on a wealth of experience . . . I can well remember the intense concentration. One lived with the structure. I am tempted to say one ate, slept and shaved with it. Finally, after six months or longer, and if one were fortunate, everything suddenly clicked into place . . . The successful analysis of one structure often leads to a quite new understanding of a whole range of forms . . . It is all the more important that really key structures should be chosen for examination, as likely ventures in which it is justifiable to sink one's capital."

WASTEFUL USE OF ANIMALS

What has all this got to do with kindness to laboratory animals? Just this: empirical methods are wasteful of experimental material. So long as the worst that happens is that large quantities of expensive chemicals are poured down the drain, nobody need worry except the man who pays for the chemicals; and as a matter of fact most of the 3,500 compounds which were screened in the research I mentioned above were tested *in vitro*.* But empirical methods are equally wasteful when animals are used, and in experiments on animals ethical requirements demand, even more strongly than does scientific virtuosity, that an experimenter should not be content with pedestrian empiricism but should plan his experiments in the light of a great deal of preliminary hard thinking. First he will assemble any relevant data which may already be available; secondly he will formulate tentatively, by intuition and hard thinking, a provisional hypothesis which may explain these data; thirdly he will deduce from this hypothesis some implications that can be tested by experiment; fourthly he will design experiments to test these implications, and consult

* To forestall comment I quote the following:
"Cycloserine . . . although fairly active *in vitro*, was found to be inactive in mouse and guinea-pig tuberculosis and other infections in animals. It would normally have been rejected, but owing to its virtual lack of toxicity in animals it was tried clinically and found to be highly effective in man. This raises the question as to whether the *in vivo* results in animals are any more reliable that the somewhat discredited *in vitro* technique for the assessment of the value of a new antibiotic." (*Nature*, 1957).

a statistician if he is not sure of his design being maximally efficient; and finally he will evaluate his results. Unless his intuition has been exceptionally lucky, he will then have to see what modifications of his initial explanatory hypothesis are called for.

The difference between the planning of a fully efficient experiment at one extreme, and pedestrian empiricism at the other, is like the difference between the behaviour of Köhler's ape and that of Lloyd Morgan's terrier. It is also like the difference between precision bombing of a military target and the system in which a substantial area surrounding the target is saturated with bombs.* The precision bomber hits the mark with minimum civilian damage and waste of ammunition, and a well-planned experiment uses the minimum number of laboratory animals that will yield the required information with the appropriate degree of precision.

There is some evidence that the empirical approach is characteristic of pre-adolescence: children aged 8-12 years were found to adopt it in solving a simple mechanical problem, while children aged 12-15 years adopted an insighted approach (Piaget 1953).

TACTICS

"The theory of statistics enables an experiment to be planned so that the maximum of information may be obtained from a limited number of observations, and conversely it serves to minimize the number of observations required for a given conclusion; it has thus a humane advantage in experiments which impose any degree of

* Saturation bombing intended to destroy the railway bridge at Tarascon killed between 1000 and 2000 civilians but left the bridge intact.

hardship on experimental animals. Formerly the biologist's only resource for increasing precision was to use more animals, but R. A. Fisher and his school have created a new and powerful technique which has revolutionized the inductive method in this respect. Statistical theory enables the cogency of an inductive inference to be measured. It enables the whole of the generalizable information implicit in a given set of observations to be extracted from them, together with a specification of the degree of reliance to be rested upon it." (Hume 1947a).

It is not only among biologists that a simple faith in very large samples used to be found. I well remember how, in the early days of relativity, a physicist repeated the Michelson-Morley experiment, a laborious process, no fewer than 12,500 times, and believed that he had disproved Einstein's theory. All that he had really done, it seems, was to measure the systematic error of his apparatus (*Science* 1926 ; but cf. *Rev. mod. Phys.* 1933).

In the case of biology, the use of hundreds of animals in a blindfold quest for precision is mere blundering in the dark, and the precision of an experiment could even be impaired by increasing the size of the sample, if this should entail having to take in a more heterogenous stock.

THE NULL HYPOTHESIS

Fisher's work is based on a concept which he calls "likelihood", by which is meant the probability, as calculated from a given hypothesis, that a given experimental result should be obtained.

Thus, you formulate a hypothesis and propose to test it by the result of an experiment. For this purpose you would wish to calculate deductively from your hypothesis the likelihood that your experiment would yield those observations which it has in fact yielded, so that hence you might draw a conclusion as to the probable rightness or wrongness of your hypothesis. But here a difficulty arises, and Fisher's solution of it is the foundation of his tests of significance. The difficulty in question is that usually your hypothesis will not imply any particular numerical value for the likelihood of finding a given experimental result. For instance, suppose the hypothesis to be that a given dose of an antibiotic affords protection against a given species of bacteria, and suppose that the experimental result shows a certain proportion of successes to failures. The hypothesis tells nothing about the probability of there being such a proportion of successes to failures, since it does not entail any frequency-distribution. Fisher's master stroke is to substitute for this primary hypothesis its contrary, which he calls the " null hypothesis "—that is to say, the hypothesis that the observations are to be explained as the effect not of the antibiotic but of random unidentifiable factors. This null hypothesis does provide a basis for calculating the likelihood that the observations obtained, together with any more unlikely sets of observations, should have occurred by chance ; and if it is found that the experimental results are unlikely to have been due to chance, as the null hypothesis supposes them to have been, then that hypothesis is rejected and the primary hypothesis is taken to be correspondingly likely. In other words, the result of the experiment is taken to be significant.

LEVELS OF SIGNIFICANCE

At what level of unlikelihood should the null hypothesis be rejected and the result of the experiment be accordingly treated as significant ? The choice of such a level of significance is somewhat arbitrary, but in practice it has been found satisfactory to work to a level of either 5 or 1 per cent, and it is for these levels that statistical tables are usually included in the textbooks. In other words, the result of an experiment is taken to be significant if the odds are 19 to 1 or 99 to 1, as the case may be, against that result, or any more extreme result, being due to chance. No scientific conclusion is ever certain ; every experimenter runs a calculable risk of attributing to an assigned cause what might in fact have been due to chance—*i.e.*, to the random effect of undetermined factors. How great is this risk? A man who habitually works to a level of significance of 5 per cent will make this mistake, on the average, once in every twenty experiments. One who works to 1 per cent will do so only once in every hundred experiments.

TABLE I—TEST OF AN ANTIBIOTIC AGAINST *Clostridium septicum* WITH FIFTY MICE

	Died	Survived	Total
Controls ..	25	0	25
Treated ..	1	24	25
Total ..	26	24	50

MATHEMATICS COULD HAVE SAVED MICE

Let us apply these considerations, by way of example, to one of a series of tests actually carried out on an antibiotic. The object of this particular test was to ascertain whether a certain dose would protect mice against *Clostridium septicum*, and the result is shown in table 1. Although common sense will tell us by inspection that this result was highly significant, let us calculate a numerical test of significance to see whether a smaller number of animals would have sufficed. Using Fisher's exact method* of calculation (Kendall 1943, Fisher 1944, Hume 1947b), we find that the result shown in table I is significant at a level of about 1 in 5,000,000,000,000. In other words, if the experimenter were to repeat his experiment, which took 10 days, continuously once during each subsequent 10 days, he might expect to make a mistake, by attributing to the antibiotic what could also have been due to chance, once in every 135,000,000,000 years approximately.

Now let us suppose that ten mice had been used instead of fifty. The odds are 4 : 1 that the result would have been as shown at (A) in table II, though it might have been as shown at (B). In the first case it would have been significant at a level of about 0.4% (P=0.004), and in the second case at about $2\frac{1}{2}$% (P=0.025). If, then, an experimenter continued to repeat the test every 10 days, how often would he be likely, on the average, to attribute to the antibiotic an effect which might also be due to chance? Allowing for the 4 : 1 probability mentioned above, we find that with ten mice per sample he would risk making this mistake once every $3\frac{1}{2}$ years. With twelve mice he would make it once every $10\frac{1}{2}$ years.

* The formula is as follows, the numerals in the numerator being taken from the marginal totals and those in the denominator from the grand total and the observed frequencies :

$$P = \frac{26! \times 24! \times 25! \times 25!}{50! \times 24! \times 1! \times 0! \times 25!} = \frac{26! \times 25!}{50!} = 2 \times 10^{-13}$$

where $0! = 1! = 1$, and the values of the remaining factorials are given in Fisher and Yates's table XXX. If the smallest frequency had not been zero, it would have been necessary to take the following further steps. Reduce the smallest frequency by unity, and so adjust the remaining frequencies that the marginal totals are unchanged. Recalculate the likelihood for this new 2 × 2 table. Repeat this operation, reducing the smallest frequency each time by unity and keeping the marginal totals unchanged, until the smallest frequency becomes zero. The fiducial probability P for the original 2 × 2 table is then equal to the sum of all the likelihoods so obtained, one for each 2 × 2 table.

TABLE II—THE SAME TEST WITH TEN MICE, IF THE ONE FAILURE WAS (A) LATE OR (B) EARLY

	(A)			(B)		
	Died	Survived	Total	Died	Survived	Total
Controls	5	0	5	5	0	5
Treated	0	5	5	1	4	5
Totals	5	5	10	6	4	10

Thus about four-fifths of the mice used in this test were unnecessary. The test was one of a series giving similarly uniform responses, with results of roughly the same order of significance. Moreover death from gas gangrene is not pleasant. But it is easy to guess how the use of this excessive number of animals must have come about. The conventional way of testing such tables is to use χ^2, and the published tables of χ^2 would entirely fail to reveal the wastage that was occurring. Unfortunately Fisher's exact test is but little used, though the computation it entails is not laborious in cases of this sort if the table of factorials be used. When we bear in mind the enormous number of mice used every year, we can guess at the humanitarian advantage that might accrue from an extended use of the tools which statisticians have put into the hands of biologists.

CHOICE OF SAMPLE-SIZE

How can an experimenter decide, before carrying out an experiment, how many animals he should use if he is to avoid wasting them on meaningless precision, and yet be pretty sure of obtaining a significant result if the effect he is looking for really exists? Usually an experiment does not stand alone but forms part of a series; for instance, the experiment cited above formed part of a series of similar tests on the same antibiotic. In such a case experience can enable a roughly suitable decision to be made. A. H. J. Baines has given an exact solution for the particular case of a variance ratio (Hume 1947c), but the whole problem has not, so far as I am aware, been fully explored by statisticians.

Economy in sampling might be achieved at any desired level of significance by the method of serial sampling or sequential analysis, but hitherto this has not been extensively tried out in biology (Finney 1955).

THE ANALYSIS OF VARIANCE

Precision can be greatly increased, and the requisite number of animals per sample correspondingly reduced, by an appropriate use of the analysis of variance. For instance: "Up to fivefold or sixfold increases in precision have been reported as a result of segregating differences between individual animals in assays of coal-tar antipyretics in cats and parathyroid extracts in dogs. Twofold to threefold increases in precision as the result of the elimination of differences between litters in the assay of vitamin D in rats were indicated by Bliss." (Emmens 1948a). The simplest design of this kind is the randomized block (Fisher 1942a, Finney 1955).

FACTORIAL DESIGN

The factorial design of experiments, developed by Sir Ronald Fisher (1942a) and further elaborated by F. Yates and others primarily for use in agronomy, has been applied also in the laboratory. It has the humanitarian advantage that, if properly used, it greatly reduces the number of animals required for producing a given amount of information, and it has the scientific advantages of indicating the interactions between different treatments and of providing a broad basis for inductive inference. It differs from earlier experimental methods in that, instead of a single factor being varied while all other conditions are kept constant, two or more factors are varied simultaneously, subject to certain precautions against the confounding of the effects of any factor with those of other factors. The factors may be treatments, such as different drugs given at various dose levels, or they may be such things as litters, strains, cages, or other environmental conditions.

To avoid confounding the effects of different factors, each level of every factor must be combined with each level of every other factor. If, however, all possible combinations of factors were made—if each level of every factor were to be combined with every possible *combination* of the various levels of the other factors—then, instead of an economy in sampling being effected, the number of observations entailed would increase rapidly as the number of factors increased, and would quickly become enormous. Fisher has overcome this difficulty by using Latin and Græco-Latin squares. By this device one or more of the interactions between factors are confounded with the error and are sacrificed, but the main effects need never be confounded. Thus at least as much information is got as with the one-factor method, and usually more information, and meanwhile the requisite number of observations is greatly reduced.

Economy can be effected by doing a factorially designed experiment with a single Latin square in the first instance. The result may be conclusive, but if not—*e.g.*, if it yields a level of significance of 7 or 10 per cent—the experiment can be repeated with a different Latin square of the same set, and if necessary with further such squares, until by combining the results a decision can be reached. In other words, it is possible to adopt a sort of sequential sampling by means of successive Latin squares, and stop when the combined results become decisive.

A COMMON FALLACY

It seems desirable to call attention here to a common fallacy which arises from the obvious fact that results obtained with a single homogeneous sample may, for all we know, be peculiar to the strain from which the sample was taken. The fallacy consists in supposing that in order to obtain a broad inductive basis a heterogeneous stock should be used. It would be as if you were to estimate the value of a pocketful of silver by counting the coins as coins, without sorting the sixpences, shillings and half-crowns. The proper procedure is, of course, to use several different homogeneous samples, by using a plurality of pure lines (or preferably Fl crossbreeds), and to allow for the variance between samples ; for otherwise the experimenter deprives himself of the possibility of making a relatively precise estimate of the error (Fisher 1942b).

MISSING OBSERVATIONS

Now what happens to a factorial experiment if one or more of the observations are lost ? Suppose, for instance, that one of the animals is found to be suffering severe pain, so that it becomes an obligation of honour, as well as of humanity, that it should be painlessly killed in accordance with the terms on which a Home Office licence has been accepted. Is the whole experiment lost ? No, for the whole of the information provided by the remaining observations can be recovered by means of the principle of least squares (Fisher 1942c, Emmens 1948b, Wright 1956) and the only information lost is then the information which would have been contributed by the lost observation itself.

TWO QUESTIONS

In relation to the methods of estimating quantities by means of regressions—in bioassay for instance—I should like to invite consideration of two questions.

(1) A great many assays depend on the determination of an ED50 or an LD50—that is, on a quantal response which entails the counting of all-or-nothing events (deaths and survivals). That quantal methods are statistically inferior to those which use a continuous variate is recognised by statisticians (Emmens 1948c). It is technically preferable, therefore, to use wherever possible a continuous variate such as body-temperature, a reaction-time, the weight of body or organs, the pulse-rate, or an analysis of blood or urine, rather than a discontinuous variate such as a count of deaths and survivals; and meanwhile, from a humane point of view, it is desirable to avoid using death as the end-point if some pleasanter technique can be found. One cannot help wondering how far the extensive use of the 50%-survival test is a hangover due to habit and custom, and whether suitable continuous variates have been sought as diligently as could be desired. Even for testing toxicity with an LD50, death might not be the only possible end-point that could be chosen if the phenomena of the moribund state were to be adequately analysed.

(2) At what level of precision should an assay be made? A drug will usually be assayed with a view to its use in clinical practice, and in clinical practice the response to a given dose varies greatly from patient to patient. While a few glasses of port may put one person under the table, the more robust guest may be a three-bottle man. In other words, the clinical response to a drug may be expected to vary very widely between patients. Is there any point in assaying batches of pharmaceutical products with a precision high enough to be swamped by the variance of the clinical

response which will be found in practice when the product is put to the uses for which it has been manufactured?

Let the prescribed dose of a given drug be P; let the optimum dose for a given patient be D; and let the dose actually given be G. The error in the dose given is (D-G), the error in the prescription—*i.e.*, the deviation of the prescribed dose from the dose which would be optimal for that particular patient—is (D-P), and the error in the assay is (P-G). Now

$$\text{var}(D\text{-}G) = \text{var}(D\text{-}P) + \text{var}(P\text{-}G);$$

thus if the response varies greatly from patient to patient, so that var(D-P) is large, any minute precision in the assay, making var(P-G) small, will be swamped by the variance of the patient's response, and no advantage to the patient will accrue from it. For example, if the standard error of the prescription is 10 mg. and that of the assay is 1 mg., doubling the error of the assay will increase the error of the dose given by only $1\frac{1}{2}$ per cent, the variances being the squares of the standard errors.

The question arises, therefore, whether some of the standards laid down by regulation under the Therapeutic Substances Act are not unrealistic; whether, in fact, a proportion of the animals used in that connection are not wasted.

REFERENCES

Blackman, V. H. (1956) *J. exp. Biol.* 7, 9.
Bragg, L. (1948) *advanc. Sci., Lond.* 4, 168.
Emmens, C. W. (1948a) Principles of Biological Assay; § 19.4. London.
— (1948b) *ibid.* § 9.4.
— (1948c) *ibid.* § 19.1.
Finney, D. J. (1955) Experimental Design and its Statistical Basis; § 4.8. London.
Fisher, R. A. (1942a) The Design of Experiments; ch. 6. Edinburgh.
— (1942b) *ibid.* § 39.
— (1942c) *ibid.* § 58.1.
— (1944) Statistical Methods for Research Workers; p. 96. Edinburgh.

Hume, C. W. (1947a) *in* UFAW Handbook on the Care and Management of Laboratory Animals, first edition, p. 286.
— (1947b) *ibid.* p. 315.
— (1947c) *ibid.* p. 343.
Kendall, M. G. (1943) The Advanced Theory of Statistics ; vol. 1, p. 304. London.
Köhler, W. (1925) The Mentality of Apes ; p. 132. New York.
Nature, Lond. (1953) 172, 322, 323.
— (1957) 179.
Piaget, J. (1953) Logic and Psychology ; p. 19. Manchester.
Review of Modern Physiology (1933) 5, 203.
Science (1926) 63, 105, 436.
Tippett, L. H. C. (1941) The Methods of Statistics. London.
Woodworth, R. S., Marquis, D. G. (1949) Psychology ; p. 493. New York.
Wright, G. M. (1956) *Nature, Lond.* 178, 1481.

The Religious Attitude Towards Animals

Read at the World Congress of Faiths, 1951, and reprinted from *The Hibbert Journal.*

ANTHROPOMORPHISM

THERE ARE two subjects in connection with which sceptics are especially apt to raise objections based on the charge of anthropomorphism; first religious beliefs, and secondly beliefs about the minds of animals. In dealing with these two subjects in conjunction with one another it seems appropriate, therefore, to begin by discussing this question of anthropomorphism.

I should be the last to disclaim anthropomorphic modes of thought. To believe in God is to believe at least that the universe has a purpose, and the notion of purpose is got by analogy with the human purposes of which we are conscious in our own minds. Similarly, when we deplore cruelty to animals we are imputing to animals feelings of pain or fear analogous to the feelings which we ourselves experience. Thus it is perfectly true that our thought on these subjects is anthropomorphic. What is overlooked, however, by those who regard such modes of thought as invalid, is that all human thought about reality is necessarily anthropomorphic. Even the concept of cause and effect, which is as indispensable in common life as it, or some equivalent category, is in scientific investigation, is got by analogy with the action of the human will. To recommend that we should drop anthropomorphism is like asking us to jump out of our skins.

Are we to conclude, therefore, that all human thinking is invalid? On the contrary, anthropomorphic modes of thought are perfectly justifiable because the human mind is a part of, and is adapted to, the universe in which it has to live. Man is akin to the lower organisms through a continuous process of evolution, and this fact justifies us in applying the human analogy, with all due caution, of course, in interpreting their behaviour. Again, if man is made in the image of his Creator, as the Jewish scriptures assert and as we may infer from the fact that the universe is at all intelligible to man, then human analogy is a perfectly valid mode of thought in theology, within its proper limits. Of necessity every analogy must fail when it is pushed too far. We should be thinking amiss if we attributed to the Deity the pettinesses of human nature, and we should equally be thinking amiss if we attributed to the lower animals intellectual powers of which they give no sign. But that is merely to say that our analogies must be applied intelligently and with a due regard to observable facts. I shall, then, in what follows, write as an anthropomorphist who regards all forms of positivism as the wearing of epistemological blinkers.

REVERENCE AND RESPECT

The two outstanding characteristics of a religious attitude of mind are reverence and what is traditionally called " love ". Love for a fellow-creature might in modern English be better, perhaps, called " kindness " ; this word is reserved for an attitude towards creatures as distinct from an attitude towards their Creator. On the other hand, the word " reverence " is generally reserved for the Deity and the corresponding word " respect " is applied to creatures, but these two attitudes are correlated : " *Est enim pietas*," says Cicero, "*justitia adversum deos*." Let us now discuss reverence or respect and defer the discussion of kindness for the moment.

A truly religious man will feel respect for all living creatures. A good example is found in the traditional Jewish ritual slaughter. (I may remark in passing that although schechita as practised to-day is not as painless as modern methods of slaughter, since the use of preliminary electric stunning is at present forbidden, it does embody rules having a humane intention—the knife must be sharp and there must be no sawing—and when first introduced it was vastly superior in respect of humaneness to the methods of slaughter then prevalent among Gentiles. But that is a digression, for we are now discussing respect for animals, or reverence for life, which is quite a different thing from kindness.) The system of schechita recognizes that the taking of life is a serious matter, not to be lightly esteemed. Jewish slaughter is carried out by shochetim who are specially appointed and trained and do their work with a certain ritual and with introductory prayers.

In the western world all too little respect has been felt for most of the lower animals. The objections urged against Darwin's hypothesis on its first appearance arose not so much from any genuinely religious feeling as from Victorian repugnance to the idea of being related to vulgar apes and monkeys; witness Bishop Wilberforce's famous

challenge which drew a well-deserved rebuke from Huxley at the Oxford meeting of the British Association. On this occasion Huxley was the more truly religious of the two. It is from reverence towards God that a Jesuit teacher, Father Martindale, would deduce a right attitude towards animals. At a symposium organized in 1928 by the University of London Animal Welfare Society he said:

"If one does believe in God, I think one is foolish if one does not begin from so vast and profound and rich a certainty, and deduce lesser laws and ideals therefrom. Hence I find myself from the outset possessed by a reverence for all that reveals God to me . . . Since, then, I perceive God, the maximum of Existence and of Agency, existing and acting and making, I am filled with horror at the thought of being the opposite of any of that. The negative, the inert, the destructive, make me shudder, as being unlike God."

Father Martindale then spoke of reverence for flowers, and went on:

"Still more lowly and reverently must I order myself towards animals, but then I must be logical and sincere, and not do so only towards animals that I like, or that like me, or that are lovely to look at, or that are large."

The reference to animals that are "lovely to look at" reminds me that Charles Kingsley, while in procession from vestry to chancel, once noticed a beautiful butterfly on the ground and stopped the procession in order to rescue the insect, whereas if at an earlier stage of its career it had fallen into Kingsley's hands as a caterpillar it would have run a serious risk of being impaled on a hook by that indefatigable angler. The Committee on Cruelty to Wild Animals which recently reported to the Home Secretary noted that almost all the sentimental concern of the British public is directed towards animals that are regarded as beautiful or attractive, such as foxes and deer, and few people seem to be in the least concerned about what happens to rats. This sentimental discrimination is clearly irreligious, because it is shallow and self-centred. Moreover it is irresponsible: the dog, which is almost worshipped in Britain, is regarded with extreme disrespect by Mohammedans; the cow, which is sacred in India, is a mere milk-factory in Britain; and so on. This partiality ought not to be. If our respect for animals is to be founded on a principle, it ought to be universal and impartial.

Reverence for life, and respect for animals, is a strong characteristic of Hindu and Buddhist theology and reaches its limit among the Jains, who will not even kill an insect. Exaggerations apart, there can be no doubt, I suggest, but that in recognizing the kinship of all living creatures these religions are on sure ground and have rendered a great service to religious thought. On the other hand reverence for life, when it is carried to the point where a man will allow an animal to die of starvation rather than put it out of its misery, impinges on a quite different principle, that of kindness, to which we must now turn.

LOVE OR KINDNESS

As the late Professor Harold Smith has pointed out*, a kind and friendly attitude to animals has always been inculcated by the Jewish religion and is also implied in the New Testament. It is also embodied in the doctrine of Ahimsa which plays an important part in Hindu ethics, and in Buddhist theology which teaches that to ill-treat animals is to ill-treat oneself, all being regarded as manifestations of the Absolute. But when we ask how far these religious principles have passed into practice we find an appalling degree of inconsistency. In India the sufferings

* *The Influence of Religions on Man's Attitude towards Animals*, by Professor F. Harold Smith (UFAW).

of the water buffaloes are notorious, and every traveller tells the same tale of prevailing callousness and apparently heartless cruelty. In China things are even worse. All reports from travellers seem to indicate that there is appalling cruelty throughout the East, in defiance of the principles which have been inculcated by Eastern religious teachers.

Professor Smith also pointed out that although humane principles are inculcated or implied in the New Testament and in the Koran, most Christian and Mohammedan theologians have failed to follow up the implications of these principles, a fact which he attributes to the influence of Aristotle. This neglect of our duty to animals by Christian theology has to be admitted with shame and profound regret. In 1947, however, the General Assembly of the Church of Scotland put on record its recognition of this duty, and its example was followed in 1951 by the national Methodist Conference. Roman Catholic theologians are, through St. Thomas Aquinas, more influenced than others by Aristotle, and it is they who have most difficulty in shaping a theoretical expression of the humane implications of their faith, being hampered by abstract logical puzzles about the nature of rights. This fact, and the appalling cruelty which prevails in most Roman Catholic countries, notably Spain and South America, have told heavily against the prestige of the Roman Catholic Church in animal-loving countries like our own. We must therefore applaud the gallant efforts of the struggling little Catholic Study Circle for Animal Welfare, which has obtained from the Holy Office a ruling that it is sinful to torture dumb animals and that such sins are degrading to the soul and disposition of the tormentor.

Mankind is notoriously indisposed to carry theory into practice. We have seen that in Eastern countries, where the prevailing theologies inculcate consideration for animals, the most shocking cruelty prevails in practice, and yet correspondingly in Christendom, where theologians have all but ignored the subject, many hundreds if not thousands of saints have shamed the theologians by giving practical effect to the spirit of their Master. " The souls of the saints are exceedingly gentle " said St. Chrysostom " and loving to man, and they show their kindness even to brute beasts." Some examples have been collected in *The Church and Kindness to Animals*, published by Burns and Oates, and in Mrs. Armel O'Connor's *Saints and Animals*.

But it is in the Protestant countries of north-west Europe that the duty of kindness to animals has been embodied in effective legislation and seriously studied as a branch of jurisprudence. The curious fact is that the movement for the protection of animals originated in the Evangelical revival of the eighteenth century, which was essentially other-worldly and primarily concerned with the saving of men's souls. Whether this was another example of human inconsistency, or whether there was an underlying logical connection, I shall not discuss.

Even in English-speaking countries, however, there is much real cruelty. In Britain between 30 million and 40 million animals are crucified in gin traps every year*, and in Australasia twice as many. On marginal land in Australia tens of thousands of cattle are left to die slowly of thirst and starvation in the drought every few years, for the sake of the profits accruing in the intervening years when the gamble on rainfall comes off. But I will not prolong this list of horrors, particulars of which can be obtained, by those interested, from responsible animal-protection societies like UFAW or the R.S.P.C.A., or the Scottish or the Ulster S.P.C.A.

* This paper was written before July, 1958, when the gin trap became illegal in England and Wales, and partly so in Scotland.

THE PROBLEM OF PAIN

Every theist must have felt baffled and perplexed by the existence of pain. Both Christianity and Buddhism have faced it bravely, and they have sought solutions in two diametrically opposite directions. Buddhism, which more explicitly recognizes the specially baffling fact of pain among animals, hopes eventually to extinguish suffering by extinguishing desire. Christianity, on the other hand, with the Cross as its symbol, looks for a triumph through suffering, in which the redeemed shall " have life and have it more abundantly ", and implies that for the purpose of redeeming the universe God shares in its suffering instead of standing aloof. " The whole creation " says St. Paul, " groaneth and travaileth in pain together until now ", and he holds that " the creature also shall be delivered from the bondage of corruption into the glorious liberty of the children of God ".

Nevertheless, the problem remains : the fact of pain, and especially the undeserved suffering of animals, appears to contradict the deliverance of the religious consciousness that God is good. Let us first get our facts right. On one hand it would be dishonest to evade the difficulty by pretending that animals cannot feel, because obviously they can and at times they undergo horrible suffering, especially at the hand of man. On the other hand we must not exaggerate the proportion which suffering bears to happiness, as we should do if we were to imagine, like J. S. Mill in his *Essay on Nature*, how we should feel if we were to perceive all the suffering in the world simultaneously. This would give a biased impression, for we should be guilty of selective sampling and missing out the concomitant happiness. If an army nurse were to base her whole picture of a soldier's life on her experiences in a Casualty Clearing Station during a battle, her picture would be false. Wild animals mostly come to our attention when they are being killed, but in fact only a small percentage of their lives is spent in dying. They live rough, but they live with zest. Infant mortality is high in the wild, but for mature and maturing animals, at all events, the zest of life would seem to predominate substantially over its pain. Moreover, pain is limited in its duration by death and in its intensity possibly by swooning, even among animals, though this is not certain. J. Crowther Hirst once collected evidence from a number of persons who had been mauled by lions or tigers, and all said that in the excitement they felt no pain. This is interesting, though it would be rash to rest too much weight on it.

With this factual background we return to our problem. One solution that has been proposed is the Hindu doctrine of Karma, but to me this seems to be open to a serious objection which will be mentioned later. From the Christian point of view the great Bishop Butler put forward two suggestions. In the first place he made much of what he called the " constitution and course of Nature ", the necessity for fixed natural laws. He would assuredly have welcomed the doctrine of natural selection, which carried his idea forward in detail. Nevertheless, if in this way we have advanced a little further down the tunnel there is still no glimpse of the distant end. Bishop Butler's other solution was the same as that given in the book of Job, an appeal to human ignorance; we know enough, he said, to justify theism at least as a working hypothesis, but not enough to justify the objections against it. I want to pursue this point of view a little further in relation to the apparent contradiction, arising out of the sufferings of animals, between belief in the omnipotence of the Creator and belief in His benevolence.

Most of us make our first acquaintance with the art of

reasoning through geometry, in which a contradiction infallibly proves the untruth of the hypothesis that implies it. But geometry traces the relations between concepts which are adequate and final, being completely determined by definition, and for this finality they pay by being imperfectly adequate to reality. What Euclid tells us about perfect circles is infallibly true, but then there are no perfect circles in Nature, if only because matter is discontinuous and because space is not Euclidean. In physical science, on the other hand, our main task is to frame a conceptual system which corresponds as adequately as possible to experience and therefore to reality, and we can do this only by successive approximations; our concepts are never fully adequate. A contradiction proves that our theoretical concepts are inadequate as representations of reality, but not necessarily that they are radically false.

In geometry, then, a contradiction proves the underlying hypothesis to be false. In physical science, on the other hand, a contradiction often merely marks a growing-point of knowledge. Fifty years ago physicists were perplexed by a flat contradiction with regard to the distribution of energy in the spectrum, but the application of thermodynamical concepts to radiation was not thereupon rejected as untenable. In due course these concepts were modified and the contradiction was resolved by the introduction of the quantum theory.

Theodicy is still in the stage in which a contradiction persists between our concept of the power of God and our concept of the benevolence of God. This means that one or both concepts are inadequate, but that is not to say they may not be the truest we are at present capable of forming. A similar contradiction long persisted between the concept of light as consisting of waves and the concept of light as consisting of particles; both are true, but they are irreconcilable unless they are modified or developed in accordance with quantum mechanics. Thus for most people, who cannot understand quantum mechanics, the contradiction will always persist.

Some have sought to resolve the contradiction in theodicy by modifying the concept of God's omnipotence. Bishop Butler, for instance, suggested that there may be " impossibilities in the nature of things " of which we are ignorant. David Hume and J. S. Mill postulated a powerful but not literally omnipotent Creator. Others have sought to modify our concept of God's love, for instance by emphasizing the value of discipline and free will, and pointing to their necessary consequences. These speculations are helpful as suggesting the class of ideas into which a satisfactory solution of the problem of theodicy might fall if we were capable of conceiving of it, but they do not in themselves constitute a satisfactory solution, and we must content ourselves with remembering the difference between a contradiction in geometry, which proves a thesis to be untrue, and a contradiction relating to reality, which may merely arise from our concepts being inadequate ; these may be the truest we are at present capable of forming.

A FUTURE LIFE

In the Western world it has generally been supposed that there is no future life for animals, but in the Eastern world there is a prevailing belief in reincarnation, with the condition that every creature's weal or woe depends on its conduct in a previous existence. This theory is open, however, to a strong objection arising out of the nature of personal identity. What do I mean when I say I am the same person now that I was half an hour ago, or a year ago, or when I was a boy ? I may mean one or more of three things. First, there is continuity ; we say that a tree is the same tree to-day as it was twenty years ago, because each state of that tree has followed continuously upon an almost identical preceding state. Secondly, personal identity may be said to reside in the fact of consciousness. My present is linked to my past by memory, and it is this continuing consciousness which, according to John Locke, constitutes personal identity. John Locke's view was challenged by Bishop Butler, who argued that consciousness is only relevant as the act of perceiving an identity which is always there, whether we are conscious of it or not. What this amounts to is, I think, that personal identity consists in one's characteristics to-day closely resembling what they were previously. Now in the case of reincarnation none of these three criteria applies. There is no continuity in the transition from man to animal or *vice versa* ; there is no memory of former life ; and there is no resemblance between a man's characteristics to-day and those of some animal which he is supposed to have been in a previous existence.

And here I should like to interpose a curious speculation. Suppose that an animal should die to-day and that next year an animal of the same species with an identical temperament, with similar genes in its cells, were to come into existence. Should we say that it was the same or a different animal? It would not satisfy either of the first two criteria of personal identity, but it would satisfy the third.

I now come to the Jewish, Mohammedan and Christian view of immortality, which implies not reincarnation into the present world but a future existence in an entirely different state. The only justification which I am aware of for believing this doctrine derives from what we may believe as to the character of God. This is certainly true of any belief based on revelation ; and independently of revelation it is felt that God must see the lovableness of those whom we love, and must wish to preserve them (with the corollary of conditional immortality, *viz.*, that those who are irredeemably evil may eventually be extinguished). This view has been ably argued by Canon Streeter and his collaborators in their symposium entitled *Immortality*, and it obviously applies to animals as well as men. According to one of Canon Streeter's collaborators, Clutton-Brock, one cause of the prevailing disbelief in a future life " is the strange assertion . . . that animals have no souls. This did not matter so long as men saw no likeness between themselves and animals ; but, now that a

thousand discovered facts have proved the likeness, the contention is obvious that since animals have no souls men can have none either, and must die like dogs. But how if dogs die like men ? How if animals are like men rather than men like animals ? Perhaps the last piece of Christian humility that we have to learn, with St. Francis, is that the blackbeetle is our brother. Perhaps it is the generic snobbery of man, more than anything else, that has deprived him of his highest hopes ".

Bullfighting in France

Reprinted from *The UFAW Courier* No. 17, Autumn, 1960.

DURING AND after the recent Santo Estello of the Felibrige I gleaned new information about the above, and learned that (1) the political implications go much deeper than we realized but (2) our policy of contrasting the *corrida* with the *course aux cocardes** is the only one that offers any possibility of progress. The background is as follows.

For seven centuries Paris has sought to domineer over the Midi and to destroy its language and culture. As a result the Meridional resents intensely anything that looks like an attempt from outside to interfere with his liberties, or anything that is derogatory of his traditions. In 1854 the great poet Frederic Mistral of Maillane (the earlier cantos of whose *Mireio* are probably the finest poetry since Homer) founded the Felibrige, initially to defend the Provençal language but eventually to preserve local customs and traditions also. Politically the Felibrige tends towards home rule of a federal, not a separatist, type, but its essential task is to maintain Provençal culture in the teeth of the domineering interference of Paris.

Later a younger disciple of Mistral, the Marquis Baroncelli, founded *La Nacioun Gardiano*. He lived in the Camargue and had a herd of native black bulls whose gardians, together with a number of amateur gardians like the poet

* In *Two Ways with a Bull*, the cruel Spanish bullfight was contrasted with the traditional Provençal *course aux cocardes* in which unarmed players known as *razeteurs* have to snatch trophies from the horns of experienced bulls, maltreatment of which is strictly forbidden. No horses are used.

d'Arbaud, were devoted to the ideals of the Felibrige. Baroncelli associated the cult of the bull with that of the traditional costume of the Arlésiennes, and instituted those picturesque games in which gardians, Arlésiennes, flowers, white horses, and gallantry are the essential ingredients. He left his Palais de la Roure in Avignon to Madame Flandressies-Esperandieu, who has fitted it up as a memorial and has mounted in the courtyard a number of antique bells which she has dedicated to various Felibre and which she rings unexpectedly from time to time by remote control.

In 1921-23 a campaign against bull-fighting was organized by the Paris S.P.D.A., and a number of prosecutions were brought but, with Parisian incomprehension and that unerring instinct for doing the wrong thing which characterizes animalitarian extremists, the campaign was directed against the humane *course aux cocardes* as well as the cruel *corrida*. In 1922 the English colony in Cannes persuaded the maire to prohibit such a *course*. Here were not only Paris but the Anglo-Saxons interfering with a traditional and harmless

Meridional pastime. There was an intense reaction. The same thing had happened in 1894, and unfortunately on both occasions the Meridionals decided to defend the *corrida*—one extremism begets another—although their real

aim was to defend the *course provençale*. On 14th October, 1894, even the great poet Mistral, himself a lover of animals though he loved Provence more, had presided at a *corrida*, though he must certainly have hated it, as a demonstration against outside interference.

In November, 1921, when a prosecution of bull-fighters in Nîmes was in prospect, the Marquis Folco Baroncelli-Javon, the lawyer Bernard de Montaut-Manse, and Gaston Audry, founder of the Federation of Taurine Societies of France and Algeria, met in a stable in Le Cailar and resolved upon a *Levée des Tridents*. They were joined by Jean Grand, Captain of the Nacioun Gardiano, in launching a long manifesto to the People of the Midi and the Government of the French Republic. It recalled the war service of Provence, Languedoc, Aquitaine, Gascony, Béarn, and Catalonia, spoke of Henry IV, Mirabeau, Joffre and Foch, and claimed that " The Meridionaux have saved nothing of their customs, as a free people, except their language and their *courses de taureaux*. This language, so often threatened by culpable and clumsy Government policy, has reappeared more lovely than ever wearing the splendours of the words of Mistral. The unity of France has not been impaired by this and will not be impaired by retaining our *courses de taureaux*. For our people these games are not an empty amusement : they are the symbol of our former independence. They are our pride and glory. They remind us of all that we owe of pious affection to the memory of our forbears."

On 17th November, 1921, the day of the trial, the roads of Provence and Languedoc were thronged with gardians riding white horses and carrying tridents. From all the towns and villages of the Midi trains poured into Nîmes a crowd of patriots who assembled in the Place des Arènes and gave a tumultuous reception to a long and passionate harangue delivered in Provençal by De Montaut mounted on a white horse. He invoked the glories and liberties of the past—kings, emperors, the Consular Republic, la Reine Jeanne (" aquelo femo d'uno béuta divino "), to which somewhat mythical personality he addressed a lyrical apostrophe, and the troubadours ; he recalled the crusade against the Albigenses, Simon de Montfort, and the battle of Muret with the ruin which followed and from which " the Midi saved only its language and its *courses de taureaux* ". In the afternoon the cortège was reformed with, at its head, the député Joly, the maire, and representatives of the general and municipal councils, and delivered a petition to the government through the Prefecture. Then de Montaut, again mounted on his white horse, delivered another long harangue, referring *inter alia* to " Sweet Provence, nest of pretty girls, garden of cypresses and olive trees, paradise of light and of perfumes . . . Vivo nosto lengo ! Vivon nosti courso de biou ! Per nosti liberta, Zóu !" De Montaut then went into court to defend the bullfighters in a long, lyrical and eloquent discourse. The bullfighters were acquitted but there was an appeal to the Court of Cassation, which sent the case back for retrial to the court of Vigan, which also acquitted the bullfighters. On a second appeal the Court of Cassation remitted the case to another court with mandamus to convict. This led, however, to a vast and prolonged agitation throughout the Midi, in the press and elsewhere. F. Mistral nebout, nephew of the great poet, was very active in this, and with another young man posted a declaration in the Palais Bourbon in Paris. The movement grew rapidly in strength and wealth and attracted most of the more important members of the Felibrige, including its Queen but not the Capoulié Dr Fallen who, however, was soon deposed. Eventually, in 1951, the law was amended so as to legalize the *corrida* in places where

" an uninterrupted tradition prevails ". This is the result achieved by the lunatic policy of the S.P.D.A. of Paris, incited by a wild woman of this country, " une illuminée anglaise ".

At the annual congress, called " Santo Estello ", of the Felibrige in Toulon in 1958 I read a paper entitled " Les Provençaux Amis des Bêtes ", which ended by condemning the corrida and praising the *course aux cocardes.* This was well received, was published in the proceedings during the recent Santo Estello at Nice, and was reviewed in the local press without adverse comment. At Nice I read a paper " Vivon li Biou !" devoted entirely to this subject, and this also was applauded without criticism by those present, including F. Mistral nebout, who had been a promoter of the pro-bullfight upheaval described above. The Escolo de la Targo in Toulon, which is the liveliest section of the Felibrige, held a meeting on 8th June, which was very well attended, and I again dealt with the same subject with acceptance. The Capiscau of la Targo, M. Bachas, is also Baile of the Felibrige, and is in sympathy with us. Its Assistant Secretary, René Jonnekin (a Ch'timi from Dunquerque turned Moco in Toulon) is an enthusiastic member of the Toulon S.P.D.A. and is willing to collaborate. In Maillane on 13th June, F. Mistral nebout, who is Reire-Capoulié and the most influential elder statesman of the Felibrige, invited me to take the apéritif with him, referred me to his article published in the *Revue Fédéraliste* of September, 1923, from which most of the above information comes, and explained that what he and his friends had really sought to defend was the *Course aux Cocardes.* They had only defended the corrida as an outpost line.

The following facts are relevant: (1) the native black bulls are used *aux cocardes* only, never in the *corrida* ; (2) the *corrida* is put on by Spaniards only, not Provençaux, though many of these attend as spectators; (3) nevertheless the taurine clubs are interested in both kinds of spectacle; (4) there has always been friction between Catalans and Spaniards, and the Catalans are closely linked with the Felibrige—the loving-cup known as the "Coupo Santo", which is brought out at the annual Taulejado de la Coupo, was presented by Balaguer and symbolizes the fraternity of Catalonia and Provence. I have never met a Spanish Sòci in the Felibrige. (5) Our friend who contributed to the bull-fight number of the *UFAW Courier* (No. 7, p. 23) now holds an important judicial appointment in Languedoc, but he thinks there is little chance of successful legal proceedings anywhere at present.

Our policy must therefore be to change public opinion gradually by praising the native *course aux cocardes* and harping on the differences between it and the alien Spanish *corrida*. If we can bring the tourist agencies into this, they will be condemning the *corrida* by implication.

Vivon li Biou !

(" Vivent les taureaux !") Communication read at the Santo Estello of the Felibrige, Nice, Whit Monday, 1960.

TOUT LE monde connaît l'histoire de Jarjaye de Tarascon, qui risqua son Paradis pour aller voir une course hypothétique de taureaux dans l'abîme céleste, ayant été attiré par des chérubins qui y criaient " Té ! Li biou !" Pas moins les admirateurs de Maurin des Maures se rappellent joyeusement ce Parisien qui, ayant entendu un jeune pâtre se murmurer " An escapa ", voulait s'informer de ce que voulaient dire ces mots mystérieux et fut ébahi de voir leur effet magique, qui était de faire courir tous ses auditeurs comme des fous, ceux-ci n'ayant pas compris qu'il s'agissait de quelquechose de beaucoup moins lourde qu'un taureau de la Camargue ! Si donc je me permets de faire ici l'éloge du taureau au lieu de vous offrir un thèse strictement littéraire, je m'excuse par la considération suivante. C'est que la littérature provençale est fermement enracinée dans les moeurs et traditions du pays, et que parmi ces traditions nulle n'est plus caractéristique que l'amistanço di biou.

Celui qui a mis cela particulièrement en évidence ce fut le Marquis Baroncelli, et l'on remarque aussi avec admiration l'oeuvre poétique de la Nacioun Gardiano fondée par lui. C'est par un coup de génie qu'il a su réunir dans un seul mouvement les cultes du taureau camarguais, de la beauté des Provençales, de leur costume traditionnel, de leurs danses et de la musique des tambourinaires.

A cette heure je vous parle en Anglais qui aime et admire la Provence, et je veux poser cette question. Combien des

Anglais ont jamais vu une course provençale ? Je veux dire une course traditionnelle, aux cocardes ? Je n'en connais que deux, dont l'un est moi-même. Voilà une chose qui demande une explication, surtout vu que les Anglais qui viennent dans le Midi pour faire la connaissance de sa

culture vont en foules aux corridas qui ne sont pas traditionnelles du tout, n'ayant été introduites qu'en 1863 par l'Empereur Napoléon III pour faire plaisir à son impératrice espagnole.

Et je vous fais remarquer que la course provençale, au contraire de la corrida, a tout ce qu'il faudrait pour plaire aux Anglais, grands amateurs et des sports et des bêtes. Pour qu'un divertissement puisse être justement dénommé " sport " il est nécessaire que, comme au cricket et aux boules, les chances soient loyalement reparties entre les deux côtés, et c'est que le taureau cocardier a autant de chances qu'ont les razeteurs. La chose n'est pas du tout arrangée de façon à ce que l'équipe d'hommes gagnera toujours tandis que la bête y perdra fatalement sa vie. Le taureau cocardier peut bien remporter sa ficelle des arènes, et en tout cas il rentre chez lui sain et sauf après la lutte. (Même s'il a perdu sa ficelle il n'est pas obligé de raconter cela aux vaches.) Et puis, en fait d'amitié pour les bêtes, il est strictement interdit de maltraiter le cocardier de n'importe quelle façon. Le seul désagrément que celui-ci doive supporter c'est que quand un razeteur lui échappe en sautant légèrement au-dessus de la barrière, il éprouve une grande déception qu'il exprime quelquefois par un mugissement de mécontentement. Mais je ne demande pas qu'un razeteur se sacrifie pour consoler le taureau, puisque celui-ci peut en trouver tout de suite un autre qui convoîte ses trophées.

Les Anglais aiment les sportmen, et si vous voulez voir un vrai sportman allez trouver un bon razeteur. Il s'habille même en sportman, comme un cricketeur, et non pas comme une jolie poupée de ballet. Il doit faire voir une adresse, une souplesse, une agilité, une vitesse où l'on voit la jeunesse mâle à son plus beau. Et puis en fait de courage : le razeteur, lui, n'attend pas pour aller affronter son adversaire jusqu'à celui-ci ait été amoindri par la douleur, la fatigue et la perte de sang. Le taureau est là en toute sa vigueur native et incorrompue. Au cocardier on ne scie par les cornes, et on ne le purge pas avant le jeu " pour lui ôter son nerf ", comme dit Paquito Cantier. Il n'est pas non plus un innocent qui entre dans les arènes pour la première fois ; le plus souvent il est un as du sport, porteur d'un nom qu'il a rendu célèbre, champion qui en maintes courses a gagné de l'expérience en même temps que du renom.

Le razeteur entre dans les arènes sans aucune arme à la main. Il ne porte ni pique, ni épée, ni cape, ni muleta pour détourner les cornes pointus d'un adversaire dont la nature est de viser l'homme et non pas un drap rouge. Oui, le razeteur est digne de son noble adversaire. La Provence a le droit d'en être fière, de se vanter de cette jeunesse mâle qui avec les braves taureaux noirs, avec les galants gardians, avec leurs gentils chevaux blancs, avec la beauté et la gracieuseté des filles de Provence, avec sa campagne fleurie et ensoleillée, nous présente une contrée qui exhale la poésie comme les fleurs exhalent la fragrance.

La Provence aurait donc le droit de se vanter de ses traditions autant que de sa littérature, et cela va pour ses courses traditionnelles aux cocardes. Mais je la blâme, je blâme ses Syndicats d'Initiative, je blâme ses agences touristiques, de ne pas s'être vantés de ces courses auprès de mes compatriotes. Voilà un sport qui a toutes les qualités pour susciter l'admiration et la jouissance des Anglais, et ceux-ci n'en savent rien du tout. Une fois un des vos gardians, Denys Colomb de Daunant, a relevé momentanément un coin du rideau en nous envoyant son film " Crin Blanc ", et cela a eu un grand succès, mais à cette heure on l'a oublié et l'on n'y pense plus. Les Anglais négligent les courses traditionnelles parcequ'ils n'en ont pas entendu parler, alors que les Espagnols font en Angleterre un propagande habile et continu pour faire croire à mes compatriotes que pour apprécier un taureau il est nécessaire de le faire mourir d'une mort lente et pénible.

Donc je fais appel à tous ceux qui ont de l'influence en cette matière pour qu'ils essayent de faire comprendre aux Anglais qu'en négligeant les courses traditionnelles ceux-ci manquent l'occasion de voir des athlètes de premier ordre et de galants taureaux cocardiers faire du véritable sport ensemble, et cela d'une façon qui exprime la bonté, la force et la gaîté des traditions qui peuvent nous consoler, lors de nos brefs séjours parmi vous, des brumes et des cieux mornes qui nous affligent trop souvent chez nous.

What Rights have Animals?

Read to the Cambridge University Circle of the Newman Society, 27th November, 1961.

I MUST begin by declaring my interest. I am an Anglican, but have great respect and affection for the Roman Catholic Church.

THE BASIS OF RIGHTS

The dispute about the rights of animals is partly a dispute about the meaning of words. Many casuists deny that animals can have any rights at all, and this view has been carried into effect in French law; in France cruelty to animals is not punishable unless it is performed in public so that it might offend the feelings of a kind-hearted human onlooker. In Britain, on the other hand, animals have legal rights of their own, the infringement of which entails fines or imprisonment whether or not any human rights have been infringed at the same time. The punishment is imposed for the offence against the animal, not for offending any human being.

The theory that animals have no rights descends from the Roman jurisprudence of pre-Christian days. In Roman law only a person, *persona*, could have legal rights, and in early pagan Rome only a citizen who was father of a family could be a person; a slave was not a person, nor was a foreigner, and a paterfamilias had the right to sell or kill his children, who had no rights against him. In the course of time the privilege of personality was extended more and more widely, but this purely legal meaning of the word "person" eventually gave place, in the minds of the casuists, to a metaphysical meaning which is quite different. They say that every *intellectual* nature, with one important exception, is a person. Thus the word "person" now means something quite different from what it meant when it connoted simply the possession of legal rights, and there is no logical connection between rights and this changed meaning of the word "person". Yet the association between the two words has persisted and has been defended *ex post facto* by fine-spun dialectics.

The great Cardinal Newman had little use for verbal gymnastics of this kind. Although his Catholic soul was near to Heaven, his English feet were firmly planted on the ground. In the *Grammar of Assent* he wrote: "I am suspicious of scientific demonstrations in a question of concrete fact."[18] The starting-point of his own philosophy

of religion was conscience, and by conscience he meant not only consciousness of the moral law but also a gestalt perception of the Lawgiver implied in that experience.*

* Newman anticipated the gestalt psychology when, in illustration of this point, he wrote :— "This instinct of the mind recognizing an external Master in the dictate of conscience, and imaging the thought of Him in the definite impressions which conscience creates, is parallel to that other law of not only human but brute nature, by which the presence of unseen individual beings is discerned under the shifting shapes and colours of the visible world. . . . The new-dropped lamb recognizes each of his fellow lambkins as a whole, consisting of many parts bound up in one, and, before he is an hour old, makes experience of his and their rival personalities. And much more distinctly do the horse and dog recognize even the personality of their masters." (19)

Now conscience implies a sense of duty, and duties are correlative with rights. You can start with rights and deduce duties from them, as the pagan jurists did, but you can also start with the dictates of conscience and thence deduce rights, and that surely is the more Christian way. Let us see how the Christian conscience at its best works when it is brought to bear on man's relations with animals.

THE CHURCH LOVES ANIMALS

My friend Fr Jean Gautier, in his book *Un Prêtre se Penche sur la Vie animale*, has a chapter entitled "Does the Church love Animals?"[8] He is qualified to know the answer, being a doctor of canon law, an authority on Catholic spirituality, and Superior of the Provincial House of the Great Seminary of St. Sulpice in Paris. The conclusion he comes to is this: "The Church does love animals and has not ceased to show it. But there are in the Church ecclesiastics who do not love them."

The Church loves animals. For the first thousand years and more of Christian history the lives of the saints are full of legends of neighbourly relations with them. Some of these stories ring true: the story of St. Giles being crippled through defending his tame hind, of St. Columba with his horse, of the wild ungulates that frequented the cell of St. Theonas, for instance. There are other cases in which legends seem to have been drawn from a common stock and attached to individual saints because friendship with animals was felt to be a natural expression of the humility and charity which mark a saint. In our own day Fr Aloysius Roche has written: "Man's attitude to the brutes is elevated or degraded in strict accordance with the clearness or dimness of his spiritual vision, in strict accordance with the strength or feebleness of his spiritual capacity",[23] and Cardinal Newman wrote: "Cruelty to animals is as if a man did not love God".[20]

The present Pope is a friend of animals, and the late Pope refused the present of a luxuriously bejewelled bullfighter's cape which the Spanish bullfighting industry had offered him. Bullfighting was condemned, with severe penalties, in the papal bull *De Salute Gregis* of 1567, and this condemnation has been sustained in the Code of Canon Law of 1917.[1] But the subject has been so well discussed by Fr Jean Gautier[8] and in Dom Ambrose Agius's tract published by the Catholic Truth Society,[1] that I need not labour the point beyond citing this fact: the Holy Office has officially pronounced that animals do have some rights as against their masters or owners; that it is sinful to torture dumb animals; and that such sins are degrading to the soul and disposition of the tormentor.[1]

Attention is often called to the fact that the New Testament does not contain any such command as "Thou shalt be kind to animals". But what is often overlooked is that it also does not contain any such command as "Thou shalt not tolerate slavery". The gospel does not work in that way. It works by generating humility and charity in the minds of men who obey it, and the natural consequence of such a state of mind is consideration for inferiors.

BUT . . .

Thus the Church loves animals. How then are we to account for the fact that Roman Catholic countries are notorious for indifference to their feelings, and that in those countries any protest against cruelty is likely to be met with the retort that "animals have no souls and so don't matter"? There can be no doubt of the fact, and it is a scandal in the literal sense of that word: it is a stumbling-block in the path of humane people whose approach to Christianity is hindered by it. It is a potent weapon in the hands of the Church's enemies. But what is the reason for it? I think it is that parishioners get their view of animals from the parish priest, who gets his from the casuists, who get theirs from St. Thomas Aquinas, who got his from the pagan philosophy of Aristotle.

Now Aquinas earned the well-deserved honour of being decreed a Doctor of the Church. That means that a Catholic must treat his opinions with respect, but it does not mean that those opinions are binding on the Catholic conscience. According to the *Catholic Encyclopaedia*:—"The decree is not in any way an *ex cathedra* decision, nor does it even amount to a declaration that no error is to be found in the teaching of the Doctor."[6] Moreover, Aquinas carefully states his reasons, thereby inviting us to apply to them our own reasoning powers, to which he constantly appeals. It is not presumptuous, therefore, to scrutinize his views carefully, especially in those cases in which they are admittedly drawn from a pagan source.

AQUINAS AND ARISTOTLE

Why does Aquinas so frequently appeal to the authority of the philosopher Aristotle? One reason seems to be this. Europe had been flooded with a dangerously heretical philosophy based on Aristotle's writings in combination with neoplatonism, and derived from Aristotle's Mohammedan and Jewish commentators such as Averroës and Avicenna and Avicebron. It swept the schools and gravely imperilled the Christian religion. It was by the mighty intellect of Aquinas that the flood was stemmed, and because he had to argue with people who staked their faith on Aristotle he had to quote Aristotle against them. Moreover he was appealing to reason, and in those days reason and Aristotle meant much the same thing.

Now Aquinas took so little interest in animals that, so far as I can find, apart from a few brief and ambiguous sentences, he discussed their status only thrice in the whole of the *Summa Theologica*[2] and twice, covering the same ground, in the *Summa contra Gentiles*[3]. In two of these passages Aquinas admits that animals have souls but agrees with Aristotle that they have neither intelligence nor reason—" *non enim intelligunt neque ratiocinantur* "—and accepts his inference that they are incapable of immortality; for Aristotle had said that the mind (*nous*) with its intelligence (*theoretikes dunamis*) seems to be a species of soul, distinct from the vegetative and sensory souls postulated by him and that it " alone admits of being separated " from the body " as the immortal from the perishable ".[4] His Arabian commentators expanded this notion and Averroës inferred that the intellect is the only part of a man which is capable of immortality.[5] Aquinas rebutted the inference as to man while adopting the inference as to animals, but he really cannot have it both ways.

However, this subject need not detain us, partly because Aquinas's animal psychology is untenable in the light of modern knowledge,[10] and still more because if it is true that there is no future life for animals that fact will strengthen the moral obligation to consider their welfare in the only life they are to have.

In the other three passages Aquinas denies that animals can have any intrinsic claims upon man's compassion, and he tries to explain away any scriptural injunctions to the contrary. Again quoting Aristotle he bases this opinion on the ground that animals are "irrational". It is interesting to note that, although the Koran enjoins kindness to animals, the Arabs treat them as things, whereas the Turks, who do not inherit an Aristotelian tradition, have indigenous animal-welfare societies.

But Aquinas was not interested in animals, and his treatment of the subject was so superficial that he failed even to make the fundamental distinction between killing and hurting. Neither he nor Aristotle had any understanding of an animal's mind, which they supposed to be purely sensory. Neither of them could know that in the present century electro-encephalograms of animals would turn out to be closely analagous to those of human beings, or that several thousand scientific papers would be devoted to the psychology of the rat alone, or that the study of learning in rats would throw a great deal of light on learning in human beings.[17]

PRIDE AND SENTIMENTALITY

Now this negative teaching, which bottoms upon the pagan philosophy of Aristotle, has been adopted whole-

heartedly by some at least of the casuists, that is, the thinkers whose responsibility it is to apply moral principles to particular cases. Three factors seem to have favoured this result. One is the glorification of the intellect, and particularly of the ability to do geometry, which came into Western thought from the pagan Greek philosophers. Animals cannot do geometry, and though their intelligence is much more extensive than was formerly supposed it is much inferior to normal human intelligence. But this glorification of the intellect is pagan, not Christian. Our Lord pronounced beatitudes[13] on the meek, on those who hunger and thirst after justice, on the merciful, on the peacemakers, but not on the contemptuously intellectual. He even said, " I bless thee, O Father, Lord of heaven and earth, because thou has hidden these things from the wise and prudent, and has revealed them to little ones."[14]

Secondly, two centuries after Aquinas this glorification of the human intellect was reinforced by the humanism of the Renaissance, which tended to flatter man and almost put him in the place of God. And finally in our own days a prejudice against animal-welfare has been created by the sentimentality of all too many animal-lovers who, indeed, are more often sentimental than humane. But all good causes have their fanatics, including Christianity itself, and you would not abandon the Christian religion because there have been Donatists and Jansenists and Anabaptists. The behaviour of animal-loving cranks is a cause of but not a justification for a contemptuous attitude towards animals themselves. It affords no excuse for complacent inter-specific snobbery.

CASUISTRY ASTRAY

A particularly strong example of this contemptuousness is afforded by the late Fr Joseph Rickaby. He did indeed disapprove of cruelty practised for its own sake, but only for the self-centred reason that it is bad for one's own soul. As to cruelty which is incidental to some other purpose he wrote that " Brute beasts, not having understanding and therefore not being persons, cannot have any rights . . . They are of the number of things, which are another's ; they are chattels, or cattle. We have no duties to them ". And again—" Charity is the extension of love of ourselves to beings like ourselves, in view of our common nature . . . Our nature is not common to brute beasts but immeasurably above theirs . . . We have then no duties of charity, nor duties of any kind, to the lower animals, as neither to stocks nor stones."[22]

You will see that this position is based on two assumptions. First, the assumption that charity is a form of selfishness : " charity ", he says, " is the extension of love of ourselves to beings like ourselves in view of our common nature." Contrast this principle with our Lord's command : " If any man will come after me let him deny *himself*."[15] Indeed, it would not be difficult to show that selfishness, far from being the basis of charity or any other virtue, is at the bottom of every one of the mortal sins.

Fr Rickaby's second assumption is this, that because our nature is considered to be " immeasurably above " that of the animals, this superiority entitles us to deny them any rights, and to disclaim any moral obligation towards them. This, surely, is the mortal sin of pride in all its evil ugliness. If superiority entitles me to disclaim duties, I am at liberty to cheat or insult or maltreat anybody whose education or intellect is inferior to my own. If the lack of understanding is a bar to the possession of rights, then a new-born baby is devoid of rights, and anybody who pleases is free to maltreat it or kill it. This is the logical consequence of substituting intellectual pride for the Christian virtue of humility.

Much has been made of the statement in the book of Genesis that man is to have dominion over the animals, and dominion has been taken to justify irresponsible tyranny. But Jesus Christ revolutionized the concept of dominion, making it imply responsibility instead of tyranny. "You know," he said, "that the princes of the Gentiles lord it over them, and they that are the greater exercise power upon them. It shall not be so among you; but whosoever will be the greater among you, let him be your minister."[16]

THE EXEMPLAR

Much has been made, too, of the statement that man is made in the image of God. But man cannot create matter or souls, his existence and immortality are not independent of any other agent, nor is he omniscient or omnipresent or particularly holy. His best hope of justifying a claim to be like God is to imitate as closely as possible the incarnate Son of God.

Now if Christianity means anything it surely means this, that one who was highest in the scale of being humbled himself for the advantage of those whom he was "immeasurably above"; *qui propter nos homines et propter nostram salutem descendit de caelis*. Christians are exhorted to follow this example[21] which must, therefore, *mutatis mutandis* govern their behaviour towards creatures which are inferior to them. This is obvious, but it raises difficult questions of casuistry. What are the *mutanda*, and how are we to balance the conflicting claims of man and beast? I shall return to that subject in a moment, but first let us notice that pride is not the only mortal sin that affects the issue. The main cause of cruelty to-day is the mortal sin of avarice. Much capital is invested in it. The most striking instances are the fur-trapping industry and the whaling industry, in both of which great wealth is won by perpetrating extreme cruelty on a vast scale, but many other examples could be given. I well remember the violent opposition which the

Meat Traders' Federation offered to the introduction of humane slaughter thirty and more years ago, because of the loss of profit which they expected it to entail, and only recently the same battle has had to be fought in the United States. But it would be easy to give many other instances of cruelty arising from avarice.

A PROBLEM IN CASUISTRY

I now come back to this difficult question: Christians, in their behaviour towards creatures which are inferior to man, are bound to imitate, *mutatis mutandis*, the example set by their Master by his condescension towards beings who were so much inferior to him. But what are the *mutanda*, and how are we to balance the conflicting claims of man and beast?

The early Church was faced with a similar problem in respect of slavery. It would have been impracticable to abolish slavery overnight, though St. Paul in his Epistle to Philemon started a train of thought which eventually led to the modern reprobation of it. As to the treatment of animals, casuistry is still in a backward state owing partly to the subject's having been so much neglected by theologians and partly to the wide range of technical knowledge which it calls for. It presents such a large and difficult problem that here I can only discuss, by way of example, how it has been worked out in one particular field which I happen to be familiar with, namely, the use of animals in the laboratory.

EXEMPLI GRATIA

Two extreme views have been held. On one side the anti-vivisectionists, of whom Cardinal Manning was one, condemn all experiments on animals. At the other extreme Fr Rickaby wrote that " there is no shadow of evil resting on the practice of causing pain " provided that this is done not for the sake of causing pain but as an incidental concomitant of something else, and he instances the pursuit of science. He adds " nor are we bound to any anxious care to make this pain as little as may be. Brutes are *things* in our regard : so far as they are useful to us, they exist for us, and not for themselves ; we do right in using them unsparingly for our need and convenience, though not for wantonness ".[22] I am glad to say that these truly horrifying views are not shared by British scientists who use animals, though unfortunately they are acted upon by many in Latin countries, and in some laboratories in the United States and Eastern Canada.

Between these two extremes the truth must lic, but it is an unfortunate fact that Britain is almost the only nation to have made any serious attempt to work out the problem, which is one of the most difficult that the discipline of casuistry has to solve.

A number of human beings have volunteered to be laboratory animals on occasion. For instance, Mellanby's work on scabies was carried out on a sample of conscientious objectors who volunteered for such service in wartime. This clinical research on human beings raises ethical questions which have been discussed in a recent symposium by Sir Theodore Fox, Editor of *The Lancet*,[7] and many of his conclusions are applicable to experiments on animals, as I have shown elsewhere.[12] But animals cannot volunteer, and so somebody else must take the decision for them, thereby incurring a grave moral responsibility.

In Britain two Royal Commissions on Vivisection have laid down general principles, and the Home Office, with its Inspectors and Advisory Committee, has the duty of interpreting those principles, including what is called the " Pain Rule " ; this sets a limit to the amount of suffering that may be imposed and is attached to every Home Office certificate. One may not always agree with the decisions of the Home Office—I personally do not always do so—but they are taken with the greatest care and sincerity. Moreover we know from various sources, including a questionnaire issued by UFAW to all the biological Fellows of the Royal Society,[11] that this control by the Home Office of experiments on animals has the almost unanimous support of British scientists, among whom a humane tradition has been built up.

A SUGGESTION

Finally, as an illustration of the sort of conclusion that an amateur casuist may come to, I venture to repeat an opinion which I have published elsewhere.[9] In the first place I distinguish between killing and hurting. There is no harm in killing an animal, provided you do it painlessly, whereas in clinical research on human beings you are bound to avoid any procedure which entails a risk of death. Again, in the case of animals permanent disablement, as by hypophysectomy, can be dealt with by killing the animal painlessly at the end of the experiment, but this cannot be done with a human subject. With these reservations I suggest the following rule :—

The experimenter or inspector must put himself in imagination in the place of the animal. He must leave out of account any risk of death or permanent disablement (which alone can justify him in choosing a victim other than himself) and focus his attention on the individual experience of pain or other stress involved ; and he must now ask himself " Should I myself be willing to endure that degree of pain or other stress in order to attain the object in view ? ". If not, his decision must be negative, and no experimenter has a right to do to an animal what he would not wish done to himself, but for any risk of death or permanent disablement that might be involved.

If anybody finds this rule unacceptable I hope he will try to devise a better one, but if this is to be compatible with Christian ethics it must satisfy two conditions. First it must conform to humility, as opposed to the contemptuous arrogance which repudiates moral obligations towards inferiors. Secondly it must conform to charity as opposed to selfishness, whether simple selfishness or that extension of it which would begrudge any beneficence to species other than our own.

A CRUCIBLE FOR CHRISTIANS

In designing an experimental test a scientist tries to simplify the conditions as much as possible by eliminating irrelevant factors that might affect the phenomena to be observed. For testing the sincerity of a Christian's profession, animals offer just such a simplified situation. Being devoid of wealth, of prestige, in many species of popularity, and of various other accidents that may furnish non-altruistic motives for being beneficent to human beings, they afford material for a critical test of a Christian's humility and charity.

REFERENCES

1 Agius, Dom Ambrose : "Cruelty to Animals," p. 14 (Catholic Truth Society, 1958).
2 Aquinas, St. Thomas : *Summa Theologica* prima 76(3), secunda secundae 25(3) and 102(6, reply to objection 8).
3 — *Summa contra Gentiles*, II 82 and III 112.
4 — *ibid.*, II 58-61.
5 Aristotle : *De Anima*, II 2.
6 Catholic Encyclopaedia, art. "Doctors of the Church."
7 Fox, Sir Theodore : "The Ethics of Clinical Trials" in *Quantitative Methods in Human Pharmacology and Therapeutics* (Pergamon Press, 1959).

8 Gautier, Jean : *Un Prêtre se Penche sur la Vie Animale*, p. 107 (Flammarion, 1959).
9 Hume, C. W. See p. 133.
10 — *The Status of Animals in the Christian Religion*, Chap. 6 (UFAW, 1956).
11 — *Opinions of British Scientists on Home Office Control of Experiments on Animals* (UFAW, 1961).
12 — "Avoidance of Pain in the Laboratory," in *The Assessment of Pain in Man and Animals* (UFAW, in the press).
13 Matthew, St. : *Gospel*, v, 5-9 (Douai version).
14 — *ibid.*, xi, 25.
15 — *ibid.*, xx, 25.
16 — *ibid.*, xvi, 24.
17 Munn, Norman L. : *Handbook of Psychological Research on the Rat* (Houghton Mifflin Co., 1950).
18 Newman, Cardinal J. H. : *The Grammar of Assent*, p. 405 (Burns, Oates & Co., 1870).
19 — *ibid*, p. 107.
20 — Quoted by Dom Ambrose Agius (1).
21 Paul, St. : *Epistle to the Philippians*, ii, 5-8.
22 Rickaby, Fr Joseph : *Moral Philosophy*, pp. 248-257 (Longmans Green & Co., 1892).
23 Roche, Fr Aloysius : *These Animals of Ours* p. viii (Burns, Oates and Washbourne, 1939).

Expanding Mercy

From an address given at the 106th Annual Meeting of the Scottish S.P.C.A. in 1946.

SCOTSMEN HAVE a right to be proud of their national society, the Scottish S.P.C.A., which exists to protect those who cannot protect themselves and has, with an extraordinary economy of staff and machinery, by its beneficent work attained to a position of eminence among humanitarian bodies throughout the world. It is reasonable to hope that it will enjoy continued and increased support at a time when the poignancy of human suffering has shaken our complacency and awakened the public mind to feelings of pity.

The protection of animals is part of that general movement of social reform which has distinguished the 19th and 20th centuries. It is a mistake to separate compassion for man from compassion for animals. The two things are different applications of the same sentiment, and it is unlikely that a person who is deficient in one will be rich in the other. The way in which compassion expresses itself is different in the two cases, because the needs and nature of animals differ from those of men. For instance we try to save men from death as well as from physical or mental suffering, but often our compassion for animals will lead us to kill them, albeit painlessly; if they are starving, for instance, or if, like the rodents, they multiply at such a rate that when they are not checked they bring starvation and disease on themselves and others. But although compassion for men and compassion for animals are implemented in different ways, the objective is the same in both cases—to prevent avoidable suffering, and to confer such happiness as the nature of each species makes possible.

In the tide of social reform we can distinguish two wholly different currents. In the first place there was the self-defence, inspired by legitimate self-interest, of the oppressed classes. Sweated operatives in the cotton mills, miners, transport workers and other artisans combined to assert their just rights and to demand fair wages and fair conditions of service. Karl Marx in his philosophy of dialectical materialism appears to have assumed that self-interest is the only motive that determines the evolution of social organization, and that the self-interest of employers on one side and of workers on the other must result in a class war

without quarter. If that is true then the outlook for animals is a poor one, for they are incapable of forming trade unions or of combining in any other way for self-protection. It is not surprising that in Eastern Europe, where dialectical materialism is the prevailing philosophy, animals do not appear to be protected except in so far as they are of economic value to man.

But as a matter of historical fact there has been another factor in social reform in Britain, namely, disinterested beneficence. Sometimes the altruistic reformers have co-operated with the Trade Union movement, as Lord Shaftesbury did in promoting the Factory Acts ; moreover, the early trade-union leaders obtained their education from schools and classes provided by Christian organizations from altruistic motives. But more often the purely altruistic reformers have fought for those who could not form trade unions and could not fight for themselves. The African slaves, shipped under conditions worse than those at Belsen and Buchenwald, could never have combined to gain their freedom ; it fell to such men as William Wilberforce and Fowell Buxton to win for them the rights that were beyond their reach. Similarly Hanway sought to liberate the child chimney-sweeps who were brought up under degrading conditions and were forced, at the age of 8 or 10 years, to climb chimneys in which they were sometimes suffocated or burned alive. These reformers were not inspired by self-interest. Their motive was disinterested beneficence wrought in a spirit of service. Only by altruism of this kind was it possible to help those workers who were incapable of forming trade unions.

It is therefore not surprising that these philanthropists were also prominent in the defence of animals. Fowell Buxton was Chairman of the first meeting of the R.S.P.C.A., Wilberforce was a member of its committees, Shaftesbury used his great influence in the same direction. Conversely, the National Society for the Prevention of Cruelty to Children was founded under the inspiration of the R.S.P.C.A.

To-day there are in Britain 6½ million trade unionists [in 1961, ten million] united to defend their own and one another's personal interests, and it is recognized by everybody that but for the trade unions many workmen would be unfairly exploited. Animals have no trade unions, and nobody to defend their interests except a handful of societies among which the Scottish S.P.C.A. plays a leading part.

I have said that the stream of social reform comprised

two separate currents, the trade-union movement originally organized for self-protection, and the humanitarian movement motivated by disinterested benevolence, which sometimes helped the trade unionists but more characteristically devoted itself to those who were not strong enough to form trade unions. Now the motive behind this humanitarian movement was a Christian one. I do not suggest that professing Christians have a monopoly of humaneness or that all professing Christians are humane. What I do say is that, as a matter of historical fact, this disinterested and altruistic component in the movement for social reform was derived, sometimes indirectly but often quite directly, from the religious revival of the 18th century. Hanway, Wilberforce, Shaftesbury and General Booth, and many like them, were social reformers because they were Christians, and this motivation was deliberate and conscious.

It is by analogy reasonable to regard the protection of animals as a practical application of Christianity. Christianity teaches that it is better to give than to receive; it condemns greed, selfishness and aggression. A particularly frequent direct cause of cruelty is greed. As an outstanding instance, immense profits are made by the fur trade, whose principal weapon, the gin trap, inflicts suffering that resembles crucifixion in its details as well as its intensity. In the U.S.A. even children have been encouraged to practise trapping in order to make money. In many other cases of deliberate cruelty the motive of commercial greed enters in. You cannot be humane without some sacrifice of your pocket, and Christianity condemns covetousness.

Animals are not very intelligent, they cannot speak or write, their manners are often bad, they are, in fact, altogether lower in the social scale than ourselves. Admittedly *latrator Anubis* has many worshippers in modern Britain, as he had in ancient Egypt, but a pig or a rat, a mule or a monkey, these are usually regarded as slightly ridiculous and wholly disreputable animals. Because they are thus mere animals, social outsiders, a good many people who in principle disapprove of cruelty to them nevertheless regard it as a matter of small importance. Now this attitude is a direct denial of Christianity. The whole structure of that religion is based on the condescension of one who, being highest in the scale of being, humbled and sacrificed himself in order to serve and save those who were immeasurably his inferiors, and this is the example held up for Christians

to imitate. The Christian spirit is not being exhibited by anybody who is in the habit of feeling contempt.

Since cruelty to animals is usually motivated by greed, and apathy by snobbery, and since greed and pride, selfishness and lack of pity are a negation of Christianity, it is at first sight surprising that the Anglican Book of Common Prayer does not contain one single intentional reference of a kindly nature to animals, let alone a prayer for them. (I say "intentional" because such references do occur in psalms and lections included for other reasons.) In the *Liber Usualis* of the Roman Church, too, I have found only one such reference, in a nocturn for Christmas. Again, having made a lifelong habit of attending public worship in various denominations, I could count on my fingers the number of times I have heard such references in extempore prayer and preaching. Are we to assume that those who are responsible for public worship have been insincere in their adherence to Christian principles? I think not. Christianity as set forth in the New Testament is an organic and not a mechanical structure. The Kingdom of Heaven is compared to spreading leaven or a growing tree. When it is compared to a mechanical structure, namely a building, the structure is a growing one; it is in the course of being built. The implications of Christian doctrine are still unfolding. Seventeen centuries were needed to realize its implied condemnation of slavery. Similarly science—the spirit of free and impartial enquiry based on the love of truth—was born of Christian civilization, but not until a comparatively late date. Has not the time now come for the unfolding of Christian ethics to advance another step by a formal and official recognition of the duty, implied in the very nature of the Christian religion, of being actively merciful to all creatures that can feel?

Finally, theology divorced from its practical implications is as abstract as a course of theoretical science divorced from laboratory or field work. Jesus Christ "went about doing good", and social work such as the hospital service, education and the relief of poverty is rightly regarded as practical Christianity. It is similarly reasonable that those professing Christians who work for the considerate treatment of the humblest creatures should regard their labours as a service done in the name of their Master.

Expanding Justice

First published in 1943. Last revised in 1962.

IN ANCIENT Rome the possession of rights was originally restricted to fathers of patrician families, so that foreigners and slaves had no rights. A man could kill his son or crucify his slaves with impunity. The protection of the law was extended to wider circles of human beings gradually at first but at a rate which has rapidly increased in the modern world, though even in Pepys's time a man could still beat his wife or his maidservant at pleasure, and until 1870 anything that a married woman might earn by her own work automatically became her husband's property. It is only during the lifetime of many of us that women have achieved equal rights with men, even in Britain.

. . . even in Pepys's time . . .

As for animals, it was as a result of the movement for social reform which marked the nineteenth century that the protection of the law was conferred on a limited class of them by Martin's Act in 1822, and since then the British

Parliament has led the world in extending such legislation more widely. Progress would have been more rapid but for the fact that fair treatment of animals often entails sacrifice of commercial profits, abandonment of deeply-rooted prejudices and customs, the solution of technical problems by means of costly research, the eradication of neurotic phobias (against rodents, for instance), and the re-education of self-centred persons. Progress is also delayed by the unbalance of those animal-lovers who feel strongly but are reckless about matters of fact. Such unbalance undermines public confidence in the movement for improving the lot of animals. The reactions of human beings to animals, which seem to be deeply rooted in the unconscious (animals figure prominently in folk-lore and dreams and arouse interest, either friendly or hostile, in children and primitive peoples) present the psychologist with a field which has been insufficiently explored.

Societies for the Protection of Animals. Since much depends on British leadership in this field, it is important to maintain a strong base in these islands. In England and Wales the

. . . neurotic phobias . . .

Royal Society for the Prevention of Cruelty to Animals, in Scotland a group of societies among which the Scottish S.P.C.A. (Edinburgh) deals with the largest area, and in

Northern Ireland the Ulster S.P.C.A. promote and enforce legislation and carry out educational and numerous other practical activities. They are referred to collectively herein as the "Societies for the P.C.A.". A number of smaller societies, such as the Battersea Dogs' Home, exist for special purposes. Representatives of most of these animal-welfare societies meet quarterly for an informal exchange of views.

. . . obtain and disseminate relevant information . . .

UFAW (The Universities Federation for Animal Welfare) co-operates with these and other bodies. Its function is to solve technical problems, to obtain and disseminate relevant information, and to enlist the interest of scientists and other professional men and women and of university students. UFAW helps to compensate the harm done to the cause of animal welfare by animal-lovers of the unbalanced kind, and to form an intelligently humane body of public opinion.

Humaneness and sentimentality. Humaneness, which is concerned with the feelings of animals, must be distinguished from sentimentality, which is concerned with the feelings of human beings about animals. It is sentimental to confuse killing with hurting, and so it is to care only about popular animals, which happen to be large, useful or beautiful, at the expense of unpopular but equally sensitive animals.

Sensibility to suffering. Common sense tells us that animals feel pain, fear and distress, as humans do, though often with circumstantial differences ; to sit in a pond all night pleases a frog but would hurt a human, to bask in the sun pleases a human but would hurt a frog. The fancy that animals may be automata without feelings is devoid of any scientific basis, (see pp. 93-114).

Legal enactments. The various statutes for the protection of animals in Britain have been summarized in *The Law Relating to Animal Welfare* (UFAW, 6d.), and further information will be found in the *Legal Handbook for Inspectors* (R.S.P.C.A., 7s. 6d.).

The chief problems relating to the welfare of animals will now be summarized.

FOREIGN COUNTRIES

It is difficult for those living in the favoured countries of North-West Europe to imagine the backward treatment of animals in many countries elsewhere, especially in lands where the human standard of living is low. In the continents of Asia and Africa animal-welfare organizations are few in number, and the bulk of the domestic animals are under-fed, over-worked, neglected and maltreated; donkeys, especially, are overloaded and galled by harness. Wild animals are killed, or treated when captive, in ways which are horrifying. Those eastern religions which forbid the taking of animal life tend to encourage the preservation of unwanted animals and to hinder the humane destruction of those that are suffering, and so to foster indifference to the infliction of pain. Some even teach that suffering will benefit the sufferers in a future reincarnation. The Latin countries are notoriously backward in their treatment of animals, and even English-speaking countries other than Great Britain lag far behind the mother country. Animals overseas can be helped by educating children to respect and take an interest in them and by organizing free veterinary treatment for the animals of the poor. For such work it is sometimes impossible to raise adequate funds locally; a certain number of deserving societies with British bases or auxiliaries are known to UFAW. The Scottish S.P.C.A. publishes with its Annual Report a supplement comprising notes on overseas societies, with the title "News from Abroad". The R.S.P.C.A. has some overseas branches. UFAW has two sister societies and some representatives overseas and administers the Lorna Gascoigne Trust which makes grants for certain purposes to overseas societies.

LABORATORY ANIMALS

Supply. In Great Britain some animals, mostly rodents, are bred at laboratories; others by commercial breeders, of whom those who reach a certain standard are registered in the *Accredited List* of the Laboratory Animals Centre. The transport of monkeys for poliomyelitis serum to and

. . . in many countries . . .

through Britain has been improved but in other countries it is still unsatisfactory. Animals passing through the hands of dealers (some of whom have supplied stolen animals) are particularly liable to callous treatment. Teachers of biology who require animals for dissection are strongly urged not to have them sent alive by the dealers.

Husbandry. This has greatly improved in Britain since 1945, thanks to the *UFAW Handbook on the Care and Management of Laboratory Animals*, to the Laboratory Animals Centre, and to the Animal Technicians' Association. The International Committee on Laboratory Animals, though primarily concerned with supply, is likely to effect a gradual improvement of conditions in foreign laboratories.

Experimental treatments. In Britain the Home Office licenses experimenters and laboratories, and prescribes ethical standards : operations without anaesthetics are prohibited and the Pain Rule lays down limits to the suffering that may be inflicted. The effectiveness of any system of control must necessarily depend on the good faith and goodwill of the majority of licensees.

UFAW has a research team studying (1) replacement of animals by non-sentient material where possible ; (2) reduction of the number of animals requisite for a given degree of precision ; and (3) refinement of techniques to eliminate or reduce suffering. It has organized two highly successful symposia for experimental biologists.

In most foreign countries there is no effective legal check on callousness, and excessively cruel experiments are allowed. In the U.S.A. the Animal Welfare Institute is working in a practical way to eliminate inhumanity from research, but is subjected to unscrupulous attacks by a pressure group which claims to speak for medical science and is opposing these efforts with a reckless disregard for truth. Children are encouraged to carry out experiments on animals.

Relevant information will be found in the *UFAW Handbook on the Care and Management of Laboratory Animals* (UFAW 70s.), *Comfortable Quarters for Laboratory Animals* and *Basic Care of Experimental Animals* (Animal Welfare Institute, New York, gratis), the UFAW Symposium on *Humane*

Technique in the Laboratory (Laboratory Animals Centre, 10s., and UFAW), *The Principles of Humane Experimental Technique* (Methuen, 30s. and UFAW), the publications of the Laboratory Animals Centre, the UFAW International Symposium on *The Assessment of Pain in Man and Animals*, the annual reports of the Home Office, *Experiments on Animals in Great Britain* (UFAW, gratis), and several other pamphlets and leaflets published by UFAW.

ANAESTHETICS AND RELAXANTS

Relaxants, which are curare-like compounds, produce a deceptive appearance of anaesthesia in fully conscious animals. (The Capchur gun uses one such compound, succinyl-choline chloride.) Many new relaxants and anaesthetics are being produced. The problem is being studied by the UFAW Research Fellow at the Royal Veterinary College. Chloralose does not anaesthetize.

FARM ANIMALS

Worrying of stock by dogs. Some thousands of sheep and lambs and many poultry are cruelly savaged by dogs in Britain every year. This could be prevented by proper training and care of dogs. Owners of dogs are liable to pay damages, but the law needs to be strengthened. Insurance would worsen the situation by removing an incentive to protect stock. Local authorities have the power to require that dogs be on leads on designated roads, and this would be helpful in some districts.

Mass production of eggs, broiler fowls, calves, and, in the future, rabbits, does not necessarily entail active cruelty, though it can do so if incorrectly organized; but it can only be tolerated "providing we can stomach the thought of young animals, whether calves or only chickens, living short, overcrowded, joyless lives in order to satisfy the aesthetic whims of this feckless human generation" (R. H. Smythe, M.R.C.V.S., *Veterinary Review*, 1961). Deep-litter systems are preferable to batteries for fowls. Fowls on free range are happiest, and the initial cost and capital outlay are relatively low, but the intensive systems save labour. For calves see *Spotlight on Calves* (UFAW, gratis).

Castration and docking. Male lambs, except for the few kept for breeding, are normally slaughtered before reaching puberty, *i.e.*, before the flesh becomes tainted. It is, therefore, unnecessary to subject them to the pain of castration. The same is true of pork pigs. The odd survivor should be castrated by a veterinary surgeon with a scalpel under anaesthesia. Dieldrin has rendered the docking of lambs unnecessary, but if it is desired for breeding-stock it should be done under anaesthesia with a scalpel by a veterinary surgeon. Bull calves if kept longer than nine months and not required for breeding have to be castrated, but this should be done surgically under anaesthesia by a veterinarian.

Slaughtering. British law requires all hoofed animals to be stunned before slaughter for Gentiles. The humane-killer pistol is generally used for cattle, sheep and horses, and electric stunning for pigs. Electric stunning is humane if the electrodes are correctly applied, but unfortunately it is easier (and probably more usual) to apply them to the neck behind the pig's ears, and in this case the pig is paralysed but not stunned. Anaesthetization by means of carbon dioxide is very humane and steps are being taken for its introduction into Britain. No satisfactory method of stunning mass-produced poultry is available at present.

The handling of the animals in getting them to the killing point is also an important factor in producing or avoiding stress.

Jewish ritual slaughter is exempted from the above

statutory requirement. Great suffering was formerly entailed in the preliminary casting of the cattle, but this can be avoided by the use of special casting pens, which is now obligatory in Britain and is believed to obtain in practice. The cutting of the throat is effected with a specially sharp knife, but consciousness is believed to endure for about 15 or 30 seconds. Calves are apparently hung up by the hind legs.

The most humane methods of killing are described in *Kind Killing* (UFAW).

Markets. The Societies for the P.C.A. maintain inspectors who visit markets and do excellent work. Nevertheless in getting animals on to and off vehicles and moving them about the market there is often clumsy handling, hustling, and beating with sticks, especially by boys. The moving of unsold animals from market to market by dealers entails much stress.

Overstocking of cows. The practice of marketing cows with distended udders in the hope of impressing purchasers entails great distress and deceives nobody. UFAW urges that, for the present, Veterinary Inspectors of the Ministry of Agriculture, and inspectors appointed by Local Authorities, should use their power to require overstocked cows to be relieved as necessary by milking ; and that eventually all cows should have to be brought to market on the night before sale and milked out not more than 16 hours before sale and again immediately after.

Dehorning. Horned cattle often gore one another, especially during transport. Adult cattle are now dehorned by veterinarians, usually without pain. Calves are dehorned or disbudded without anaesthesia and this may cause pain. The breeding of polled cattle is making progress.

Difficult parturition. Selective breeding without regard to this factor has led to some breeds of cows, especially

Friesians, producing oversized calves. This results in painful and difficult parturition and not infrequently necessitates Caesarian section.

Transport between farm, market and slaughterhouse inevitably entails hardship. It could be reduced if meat were always graded on the hook instead of on the hoof. The transport of animals in Britain is subject to a number of orders made under the Diseases of Animals Acts, 1894 and 1935. These relate to such things as the supply of water and food, prevention of unnecessary suffering, carriage by rail of unfit animals, overcrowding, ventilation, and construction of vehicles. British Railways have issued for the guidance of their staff a pamphlet entitled " Customers can Complain—Cattle Can't ". The transport of animals by air is the subject of a specification published by the British Standards Institution.

Formerly very great suffering was entailed in the export from England of worn-out horses, but under an Order made in 1921 horses are now inspected by a veterinary surgeon before export, and the evil has been brought to an end. In recent years there has been heavy exportation of cattle to continental countries where the treatment of animals is backward and where very long journeys under inhumane conditions are entailed. The Government has now restricted such exportation to countries where, it is hoped, these evils are minimized.

Unfortunately the transport of animals by sea, which at best entails a good deal of stress, continues under unsatisfactory conditions between a number of foreign countries.

Foot-rot of sheep. This painful disease is widespread and causes much suffering, but can be eradicated in about a month by means of hard work. See *Foot-rot of Sheep: Stamp it Out* by Prof. W. I. B. Beveridge (UFAW, gratis). An instructional film has been prepared for UFAW.

Fly-strike. When sheep are struck by the blow-fly, maggots develop under the skin, which they proceed to eat, but it is now possible to give protection throughout the fly season by means of a single dip.

Warbles. The warble fly causes much suffering in cattle. The grub develops under the hide, where it causes an inflamed swelling and pierces a breathing-hole. After a number of weeks it works its way out through the hole, which is then of a substantial size. Extremely effective chemical treatments which involve relatively little labour have become available during the past few years. If tanners would offer a bonus on undamaged hides they might provide a sufficient incentive to effect elimination.

Worrying by flies is a real hardship to cattle and horses in the summer. The animals may be washed or sprayed with Nankor (Dow Agrochemicals) or a mixture of 1 part citronella oil, 1 of rectified oil of camphor, 20 of dimethyl phthalate, but in all cases the eyes, lips, nostrils, inner ears, and private parts should be carefully protected from the liquid.

Fluorosis. Fumes from brickworks and steelworks often contain fluorides which settle on pasture, and cattle feeding thereon for long enough break their bones and suffer much unless removed in time.

Hygiene. The prevention of disease, the proper treatment of the sick, and the practice of good husbandry can do a great deal in a positive way to avoid suffering. Avoidable suffering is very often caused through failure to call in a veterinary surgeon in sufficiently good time.

Pit ponies. In most mines these have been replaced by mechanical haulage, but in Durham and elsewhere many ponies are still used underground. While some are well cared for, others are exposed to hardships and occasionally to maltreatment.

Stabling. Bulls and stallions are often kept in confinement so close that it amounts to serious cruelty. The animals become bad-tempered and dangerous under these conditions, which are not necessary.

Slatted floors may injure cloven hoofs.

Pigs are naturally clean animals and thrive best when kept under good conditions but are not infrequently kept in dirty pig-sties. Like other creatures they are happiest on open range, but this reduces profits.

ANIMALS FOR ENTERTAINMENT

Bull-fighting: tourists. The Roman emperors offered the spectacle of martyrs being eaten alive, beast-baitings, and gladiatorial shows in which the gladiators exhibited great virtuosity, to curry favour with the mob. Bull-fights appeal to the same instincts and entail great cruelty to the bulls and horses. A prodigious and well-organized campaign is being conducted by press and radio for the purpose of attracting British tourists to patronize bull-fights in Spain. See *Two ways with a Bull* (UFAW, gratis).

Performing animals. While it is possible to train performing animals by kindness and many trainers appear to do so, cruel methods are sometimes employed. The Colvin Select Committee found in 1922 that certain charges of cruelty in

. . . naturally clean animals . . .

connection with performing animals had been established, and recommended a system for preventing such cruelty and its concealment. The showmen, however, successfully drew the teeth of the ensuing legislation. The facts have been obscured by sweeping accusations, but when all allowance has been made for reckless exaggeration there appears to be a case for strict control. Moreover performances by animals may give children a misleading impression of the nature of these creatures.

Zoological collections. The close confinement of wild species in small zoos and fun fairs and in circuses is open to strong objection.

As regards the animals in large zoological gardens, much depends on the quality of the management, the lay-out, and the species. Intelligent species may suffer from boredom and wandering species like wolves from confinement; the keeping of birds in small cages is in need of elimination. Bullying by dominant animals in a cage may amount to serious cruelty. The best modern zoos are tending to provide more adequate space, as at Whipsnade Park. The Zoological Society of London has an Animal Welfare Committee which meets quarterly and includes representatives of UFAW and the R.S.P.C.A. For the basic principles see H. Hediger, *Psychology of Animals in Zoos and Circuses* (Butterworth, 1955).

Capture and transport are the most cruel features of commercial collections, especially when capture is effected by natives of backward civilization. Mortality in transport is very high.

Field Sports. See " Wild Animals ", page 215.

DOMESTIC PETS

Caged pets. The most important source of cruelty to domestic pets is irresponsibility in the case of caged pets kept by children. Even a child who is fond of animals may forget to feed, water, entertain and clean the cage of his pet rat, rabbit, or bird. The keeping of pets is tolerable only when responsible parents accept full responsibility for their well-being. Caged pets should, as a rule, have companions of their own species.

To recommend or permit children to take away animals as pets without first ascertaining whether these will be competently cared for is a most irresponsible and reprehensible action.

Dogs. See *The Dog-Owner's Guide* (UFAW, 7s. 6d.). Although dogs enjoy most-favoured-animals treatment they may suffer from the following causes :—(1) Constant chaining-up. One remedy is to use the running chain described in a leaflet by the R.S.P.C.A. (2) Bad health due to fancy breeding with resultant physical distortion ; *e.g.*, high-bred bull-dogs suffer inevitably from respiratory troubles. Parturition is difficult in some modern breeds of dogs. (3) Lack of training is a common cause of road accidents, as well as of injury to dogs. Several organizations, such as the National Dog Owners' Association, arrange courses in dog-training, and a number of good books on this subject are available. An instructional film " Love me, Love my Dog " has been made by Barbara Woodhouse ; there is a copy in the UFAW film library. (4) Over-population. Large numbers of strays are rescued and humanely destroyed by the Societies for the P.C.A., the Battersea Dogs' Home, and other like organizations. One cause may be the irresponsible keeping of puppies for children and the subsequent abandonment of the animals. The licensable age should be 3 months.

Cats. See *The Cat-Owners' Guide* (UFAW, 5s.). Overproduction of kittens entails the problem of stray animals. It can be prevented by neutering, which is recommended for both sexes and must be done by a veterinary surgeon. Strays are humanely destroyed in advanced countries by the Societies for the P.C.A., but elsewhere are exposed to the consequences of neglect. Ear-canker troubles many cats but can easily be cured by treatment. Outbreaks of cat-stealing, carried out at night for sale to the fur trade (with no assurance of humane treatment) or to laboratories, cause distress at least to the owners of the animals.

Tortoises. Very large numbers are imported every year, and the great majority die soon from under-nourishment or immaturity. For full instructions see UFAW's leaflets and film, results of a thorough investigation.

Killing of unwanted animals. Inhumane methods are sometimes used. The correct methods are given in *Kind Killing* by Dr Vinter (UFAW, 6d.).

Treatment of sick animals. The Societies for the P.C.A. have made arrangements whereby the sick animals of the poor may have qualified veterinary treatment. Treatment by persons who are not qualified veterinary surgeons has been brought under control by the Veterinary Surgeons Act, 1948, which seems to be working fairly well.

Practical instructions for keeping pets are given by J. P. Volrath in *Animals in Schools* (UFAW, 12s. 6d.), in leaflets obtainable from the Societies for the P.C.A. and from the London Zoo, and in lecture notes with pin-up photographs published by UFAW (3d. each).

WILD ANIMALS

Fur-bearing animals. Although gin-trapping became illegal in England and Wales in 1st August, 1958, it still accounts for the major part of the cruelty inflicted in the world. About 14,000,000 muskrats are trapped annually for musquash coats in Canada, the U.S.A. and Russia, and about 25,000,000 skins of trapped rabbits go to the trade in felt hats, chiefly from Australia.

The problem of the fur industry is examined objectively by Dr F. Jean Vinter in *Facts about Furs* (UFAW, 1958, 2s.), which contains *inter alia* sections on fur-farming, furs of domestic animals, the taking of fur-bearers by means other than the gin trap, and simulated furs, together with sketch maps showing the distribution of the chief fur-bearing animals and the White List of non-trapped furs, originally compiled by Major Van der Byl and now brought up to date.

The above-mentioned White List is also printed in a leaflet which can be obtained from UFAW gratis and is suitable for free distribution.

Whales and seals. The present method of catching whales is excessively cruel. A lacerated wound is inflicted with an explosive charge, and the whale, a highly sensitive mammal, then tows a 300-ton boat for a long time, a substantial fraction of an hour, by means of a harpoon pulling on the wound. An attempt to introduce a potentially humane method (electric whaling) was catalysed by UFAW and carried on at great expense by Hector Whaling, Ltd., the General Electric Company, Ltd., and Westley Richards, Ltd. It was hampered by obstruction from the gunners and finally extinguished with the sale of western interests to the Japanese. The latter appear, however, to have become interested in the subject. It is expected that whales will be almost exterminated eventually.

Fur seals are killed in the Pribiloff Islands in an approximately humane manner, but hair seals in Newfoundland are often shot at long range and left wounded. Those clubbed on the ice are usually but not always stunned by the blow.

Trapping. The gin trap, notorious for its excessive cruelty, is now illegal in England and Wales and (except for otters and foxes) in Scotland, but see "Fur-bearing

Animals" above. Killer traps approved by the Ministry of Agriculture are still permitted in England and Wales; these are usually humane, though some ten per cent of the victims are only maimed. Cage traps are humane if regularly inspected; otherwise the victims die of thirst and starvation. Moles are killed by inhumane methods at present. Humane methods of rabbit-control are detailed in *Instructions for Dealing with Rabbits* (UFAW, gratis).

Snaring. Wire snares, which slowly strangle, are scarcely less cruel than gin traps. They are used chiefly though not exclusively for wild rabbits, and in view of their cheapness and invisibility the most effective way to get rid of snares is to get rid of rabbits.

Poisoning. Most of the poisons hitherto used for rats are excessively cruel. Especially cruel poisons are red squill, 1080, arsenic, phosphorus, barium carbonate, and strychnine. A comparatively new poison, Warfarin, is relatively humane, though not completely so in the case of wild rats; it is recommended as the best poison at present available. A Bill empowering the Home Secretary to prohibit cruel poisons became law on 5th July, 1962.

Many wild animals have been killed by seed-dressings; this matter is being watched by the R.S.P.B., the Council for Nature, and the Nature Conservancy.

Shooting. See the correspondence in the *Field* from 26th December, 1957, onwards. About 3,000,000 licences for shotguns are granted annually in Britain; applicants do not

It's all right — humans don't suffer.

have to comply with any condition other than payment of a small fee. There is no requirement analogous to the driving test for motorists. Many animals are left in a wounded condition, especially by children with airguns and unskilled shots shooting at long range or with unsuitable weapons. A large proportion of the deer in the country have been wounded with shot. Legislation is being attempted.

... analogous to the driving test ...

Big-game hunting formerly called for skill, courage and endurance, but has now been *aménagé pour le tourisme* ; et pour les incompétents.

Bounties. Bounties on squirrel tails and the like encourage unskilled and callous killing, especially on the part of children. Bounties do not appear to lead to any reduction in the abundance of a pest, and are not favoured by ecologists, The bounty on squirrel tails ended in March, 1958.

Wanton cruelty by boys. Many boys pass through a sadistic phase in which they destroy birds' nests, maltreat young birds, inflate frogs, torture hedgehogs, etc. Such cases call for either psychiatric or castigatory treatment but are all too common.

Pollution of the sea by oil. This causes a distressing death for large numbers of seabirds every year. An International Convention for the Prevention of the Pollution of the Sea by Oil has been ratified by Britain in the Oil in Navigable

Waters Act, 1955, and by some other countries including the U.S.A. It does not, however, prevent discharge of oil by ships on the high seas. Details are given in the publications of the International Committee for Bird Preservation, British Section, c/o Natural History Museum, South Kensington, S.W.7. A technical *Manual on the Avoidance of the Pollution of the Sea by Oil* is published by the Stationery Office at 1s.

Cinematograph films. Cruelty to animals shown in films, or used in making films, is illegal under the Cinematograph Films (Animals) Act, 1957. In the event of any such film being shown, a complaint should be sent to the Borough or District Council (or to the B.B.C. or I.T.V. in the case of television), and the R.S.P.C.A. and the British Board of Film Censors, 3 Soho Square, London, W.1. should be informed.

Field sports. (i) Hunting, though hard to justify in principle, affects relatively few animals and, except in the case of otter-hunting, the suffering inflicted by it is small in comparison with that entailed in incompetent shooting, for instance. See the report of the Home Office on Cruelty to Wild Animals, 1951 (Cmd. 8266, Stationery Office).

(ii) Shooting : see page 217.

(iii) Fishing. Particularly objectionable practices are the fastening on the hook of live vertebrate bait for pike ; rough and unskilful extraction from the hook ; rough handling of fish caught for return to the water, leading to sub-

It's all right — humans don't mind.

sequent infection and death; leaving on the bank of nylon lines that are liable to entangle birds by the leg.

INVERTEBRATES

It is unfortunately impossible to be consistently humane to invertebrates; in ploughing a field, for instance. But there is no justification for avoidable or unnecessary maltreatment; the fact that invertebrates can be conditioned proves that they can feel. (Charles Darwin always used dead worms for fishing.)

THE FISHING INDUSTRY

Fish and crustaceans are commonly treated as if they did not feel. This is unfortunately a mistaken assumption. The problem of inculcating a right attitude into deep-sea fishermen is one which has not hitherto been studied. For the best methods of killing crustaceans see *Humane Killing of Crabs and Lobsters* (UFAW, gratis).

PERSONAL POSSIBILITIES

University staff and undergraduates can join UFAW and enlist fellow-members of their universities.

Veterinary surgeons can make full use of their unique knowledge and opportunities, and can join the Veterinary Section of UFAW.

Teachers can arouse interest in wild life and inculcate a sense of responsibility in the care of domestic animals; co-operate with the educational work of the Societies for the P.C.A. and Protection of Birds, the School Nature Study Union, Scouts, Guides, and Young Farmers' Clubs; and join UFAW and use its educational literature and films.

Agricultural advisers and lecturers can join and collaborate with UFAW.

Farmers can keep UFAW informed of methods and devices for improving the lot of stock, and suggest problems for solution.

Biologists, physicists, chemists, engineers and lawyers can join UFAW and place their knowledge and ingenuity at its disposal for the formulation and solution of animalitarian problems.

Physiologists and pathologists can pool ideas for providing the maximum of consideration for laboratory animals, impress their views on foreign research workers, indoctrinate beginners in research, and join UFAW.

Medical men and women can join UFAW, and use their great influence with the public.

Justices of the Peace can deal sternly with persons convicted of cruelty.

Ministers of religion can use prayers for animals, stress the fact that callousness is sin, and obtain information from UFAW.

Missionaries and overseas administrators and technologists can refuse to become hardened to the cruelty they see, and try to better it by example and precept.

Artists, authors, journalists, and poets can use their talents for educating public opinion, *after ascertaining the objective*

It's all right — humans don't feel.

facts. They should ignore ill-informed and misleading propaganda.

Peers and Members of Parliament can promote animal welfare by legislation and by questions in the House.

Diplomats can acquaint foreign notabilities with the British law for the protection of animals.

Civil Servants can advise their ministers to see justice done to animals.

Policemen can keep watch against the afflictions to which animals are exposed.

Tourists can boycott bull-fights, rebuke publicists who advertise bull-fights, visit and encourage animal-protection societies in foreign countries, and make their views known when they see animals maltreated.

Cinema-goers can boycott cinemas showing any film which depicts bull-fighting or other kinds of cruelty for entertainment, and send a protest to the local Council, with a copy to the British Board of Film Censors.

Women can use all their endowments to abolish the gin-trapping for which their demand for furs is primarily responsible, and wear only simulated, or at worst White-List, furs (see p. 215).

Parents can teach children to attend meticulously to the feeding, watering, exercise and clean housing of their pets, and can avoid inducing phobias against unpopular creatures.

Children can influence parents, friends, and younger children, find out about animals by reading and by joining animal-protection societies, learn how to make their pets happy, and demand animal films at school.

All progressive citizens can inform themselves and others of the facts, join UFAW as ordinary or associate members, join the local branch of an animal-protection society, learn how to deal with lost and injured animals, buy their eggs from farms with free range, write to their Members of Parliament and local press about the needs of animals, read UFAW publications, supply UFAW with ideas, and support UFAW'S efforts.

It's all right — humans don't matter.

'what does ufaw do?'

UFAW BELIEVES —

that all animals should be treated humanely, that kindness alone is not enough and that knowledge and understanding of their needs are essential for their proper welfare. It is particularly concerned with the conditions in which animals live on farms and in laboratories, in zoos and in the wild, in schools and in the home.

UFAW INVESTIGATES—

problems such as the welfare of intensively kept poultry, the transport of livestock, alternative methods to the use of laboratory animals for the routine testing of certain drugs, humane methods of slaughter of the food animals and, where control is necessary, humane methods of culling wild animals such as seals, badgers and moles. UFAW has carried out surveys of zoos and pet shops. Detailed accounts of research and investigations are given in Annual Reports (September) and News-sheets (April) which are sent free to all UFAW supporters.

UFAW EDUCATES—

by teaching the correct methods for the care and management of animals and by disseminating accurate information on the needs of various species. It maintains contact with those who work with animals and, each year, arranges and publishes the proceedings of a major symposium on a particular aspect of animal welfare. UFAW produces comprehensive textbooks, pamphlets and information leaflets covering most animals and, by arragement, staff are available to give practical demonstrations and lectures. A full list of publications is available on request.

UFAW ADVISES AND ASSISTS—

government departments, parliamentary committees and other organisations and individuals. UFAW has many Members and Associates in the United Kingdom and Overseas and is always ready to give scientific and veterinary advice to promote the well being of animals.

UFAW SEEKS—

to achieve its aims by a professional and scientific approach rather than by engaging in emotive public controversies.

UFAW NEEDS—

donations, subscriptions and legacies. It receives no grants from Universities, Government or Commerce and relies entirely upon voluntary contributions. UFAW NEEDS MORE MEMBERS AND ASSOCIATES.

If you would like to join UFAW please write to:

UNIVERSITIES FEDERATION FOR ANIMAL WELFARE

8 Hamilton Close, South Mimms, Potters Bar, Herts EN6 3QD, England